A S T R O

T R A N S F O R M A T I O N

ASTROLOGY

Transformation & Empowerment

Adrian Ross Duncan

First published in 2002 by
Red Wheel/Weiser, LLC
York Beach, ME
With offices at:
368 Congress Street
Boston, MA 02210
www.redwheelweiser.com

Library of Congress Cataloging-in-Publication Data

Duncan, Adrian Ross.
Astrology: transformation & empowerment / Adrian Ross Duncan.
p. cm.
Includes bibliographical references.
ISBN 1-57863-262-5 (pbk.)
1. Astrology. I. Title.
BF1708.1.D86 2002
133.5—dc21 2002009850

Typeset in 10.5 Minion

Printed in Canada

TCP

09 08 07 06 05 04 03 02
8 7 6 5 4 3 2 1

Note to the reader: This book is intended as an informational guide only. The approaches and techniques described herein are not meant as a substitute for professional medical or psychological advice. Neither the publisher nor the author offers any explicit or implicit representations that using the methods in this book will bring about any particular result.

Contents

List of Illustrations

Acknowledgments

I would like to express my gratitude to all the clients over the years who have helped me gain an understanding of the all-encompassing interplay between the cosmos and the individual. I thank the students at my courses and lectures—particularly the students of AstrologSkolen—who over a seven-year period of experiential play so enriched my knowledge of astrology. No words can express my appreciation of my own teacher, Tarab Tulku, without whom I may never have discovered how empowering the right view can be.

Foreword

This book is evidence that astrology has matured beyond the anxious, mechanistic need to prove itself beyond the fool's paradise of blind faith; and, more significantly perhaps, beyond the realm of bombast, entertainment, or placation. Assuming an intelligent and psychologically tempered collective, *Astrology: Transformation and Empowerment* builds a profoundly thoughtful thesis on that foundation. This book is both a measurement of the collective development of astrology and the preparedness of the astrologer and client to receive this contemporary view.

I first came upon this book by accident—always a good sign for an intuitive like myself. I had been working and visiting my old home, and was just about to return to the United States, when Adrian Duncan mentioned to me he had a book on the go, almost done and ready to be presented to a publisher. As an author and editor, I am always enthusiastic to hear good new ideas and foster them if possible, and I sensed instantly that what he had was original and maybe even exceptional. I asked if I might please read it. I then left London for the States with the early draft of the manuscript and proceeded to read it in one stretch—with hair on end! From the very beginning, and for myriad reasons, I was eager for this book to be published.

As a 30-some-years practitioner, author, and teaching/training astrologer, I had no "practice of astrology" manual to recommend to any student—novice or graduate. Granted, we have now many brilliant in-depth psychological and archetypal astrology books covering specialist areas, but there is no single book evoking the ephemeral experience of working with clients that provides ways and means of producing the desired transformation and empowerment.

Today we have the benefit of 2,500 years of Western astrology, and in the last century, the exponential development stemming from those ancient and modern foundations. The long history of astrology has many peaks and troughs, and in the last 30 years it has enjoyed a well-earned, stupendous zenith. As a Sun-Saturn individual, I have enjoyed the peak, and in fact, I am a contributor to that peak, but felt the uneasy doom inherent in such an apex. As the fruit ripens, so it begins its decay.

I have always felt that the advent of the third millennium would mark either the suppression of astrology as we know it now or the renaissance of astrology. Thankfully, it appears that the renaissance is upon us, and as with all births, the feeding and care of the subject is now of greatest importance. Once something becomes established, and matures, it is important to take that maturity and be responsible for it.

This book is a result of all our work, and it brings the assumption of all that has gone before to its pages. Finally, we have a book dedicated to the practice of counseling astrology that looks at more than the delineation of planets, signs, and houses, or specific areas of interest such as transits or various complexes or elements of the chart. Here we have a clear guide to the principles and practice of contemporary astrology.

Through the natural evolution of ideas and the vital efforts of astrologers and scholars of astrology, we have a status. In the last 30 years (1970–2000) the research, the collation of our rich and ancient history, the popularization factor, and the flowering of contemporary astrology has created the foundation for a vigorous future. In fact, now it is time to take that place, assume it rightfully, and move on to the accepted practice of astrology in everyday life.

This book presents tools, ideas, factual cases with horoscopes, and conversations in astrology consultations. Immediately taking into consideration the problem with perception in part one, Adrian realizes that not only are we subjectively experiencing something as it is objectively happening, but also, "We reconstitute our being each instant, [and] the opportunity for change is present at all times. . . ." It is upon this premise that he bases the more therapeutic approaches to consultations, demonstrating ways of going about this transformational (and thus, empowering) process for and with the client.

The view of astrology that he gives is easily and comfortably assimilated. There is no heavy-handed philosophy here, but a philosophy based on natural science, human behavior, and the collusion between the two. The requisite explanation of astrology doesn't conflict with any single belief system, but supports a fluid, shifting reality that is both objective and subjective, not discounting the perceptions of the client, but rather aiming at the core of the perception. Once an issue or obstacle is perceived through one's own horoscope, it is entirely within one's grasp to migrate the problem toward a solution.

If we can see the root cause of our suspiciousness, we might become less so, or at the very least, appropriately suspicious rather than universally so. Similarly, with the insight gained through the chart in conjunction with our behavior, we might become appropriately hopeful, joyful, and dream-filled, rather than foolish, self-indulgent, and easily disillusioned. Adrian Duncan gives immediately useful portrayals of the skills of communication, direction, and timing in consultations. I practiced his methods on myself and was profoundly affected by the results.

All this in one book! Not only that but in part two he undertakes the context of one's perceptions and behaviors through the primary planets (personal—Sun through Mars) and beyond, to the collective framework of Jupiter through Pluto. Working with the individual in his or her collective ethos is located in context with cyclic major planetary configurations, complete with graphic ephemerides depicting the conjunctions made between Jupiter through Pluto and descriptions of the personal planetary links to those collective themes.

In part three, the focus is on the practical and therapeutic uses of various astrological themes, which inform the counseling astrologer of where to start. He advises not to wade into the most annihilating aspects but to go to three or so issues at the time of the consultation, based on transits, progressions, and the consultation chart itself.

There is no one way in this book, but as many ways as their are configurations of planets and people to live them out. You will be relieved to know that your own instincts are also part of the astrology consultation and are not dominated by planets or society. If one method of getting the client to reach the core issue is not working, then we are advised to drop it and move to a more fruitful mode! Always, the client's horoscope is the key—as this book will assure you—and the key is within the client. This book helps the novice astrologer to feel more confident and the professional more supported by the truth this book brings to our work.

Personally, I felt a greater sense of confidence in my own perceptions, I learned some new ways of perceiving myself, and I found more compassion and encouragement to change those things that I can change within myself. I feel that this book validated my work and rejuvenated my practice and outlook on the theory. It sparked new thoughts and a great deal of inspiration to carry on with my work in astrology and with people. With this book in print, in the hands of both seasoned and new astrologers, I am assured of continuing good company, new blood, and a long life of practicing the vocation I love.

It is an honor to introduce this book in a time when astrological counseling is at a breakthrough point. Our clientele now spans the full spectrum of the human experience; they live all over the planet, and the demand for credible, well-trained, and practiced astrologers will continue to grow. For astrologer, client, and psychologist, this book is an essential work.

—ERIN SULLIVAN

PART ONE

Setting the Scene

Introduction

The Philosophical Foundation of the Counseling Practice

This book is about how to empower the client and create transformation through harnessing the energies in the horoscope. These energies are shown by planetary combinations that are not simply abstract ideas, but are basic, and some might say divine, root energy vortexes with tremendous capacity for generating transformative power. However, the psychology of astrology should not be separated from its philosophy. Without a philosophical foundation on which the functioning of astrology is based, the ability for therapeutic intervention and change is diminished.

The basic tenets of this philosophy avoid the idea of causation—that the cosmos on the outside has some kind of *direct* effect on us on the inside. If we understand matter/energy, subject/object, and body/mind as two poles of the same continuum that are interdependent and interdetermining, then we can move away from the idea of the individual having no influence on his fate. If we see time as a seamless process and understand that the division of time is an artificial construct created by human minds to organize experience, then we can understand that anything is possible, now.

Astrological training is not of itself sufficient to help people who are unbalanced or traumatized, though astrologers will inevitably meet unbalanced people in their practice. There are simple techniques for helping these people along, although there is no substitute for long-term therapy for a truly disturbed client. Most of the techniques in this book assume that the client is well-balanced and not in need of being psychologically rescued. If clients are disturbed, and you don't have a training in therapy, show them the way to someone who does.

What most first-time clients want when they come to an astrologer is to see astrology work. It is a breathtaking and unforgettable experience for someone to realize that their character, behavior, and fate can be described via the horoscope. Primarily, there is a potential for the consciousness-raising realization that man and cosmos are one. That there is a bond between the individual and the solar system of which he or she is a part. It is an effective technique to induce a mild state of shock or surprise at the beginning of the consultation, as homage to Uranus, that electric planet most connected with consciousness expansion—a shock that springs from the realization of the client's connection with the cosmos. This book is about how to induce that state and what to do *after* that state has been induced. Learning to accurately describe planets, signs, aspects, and houses is an ongoing process, and through the consultation practice this learning is constantly enhanced. All astrologers have to start somewhere, and there is much to be said for taking the leap into doing consultations. The techniques in this book are intended to help those beginning this journey, and those who have been under way for some time.

Perception of Reality

It is in the nature of things that nobody can be sure of what "reality" is. What we think of as reality is a consensus of opinions that we subscribe to and are in general agreement on. Our perception of what is going on is completely dominated by our sensory apparatus, and subsequently warped by our opinions, preconceptions, and personality quirks. It may initially be difficult to accept, but what we think of as going on outside our bodies, and even inside them, is a complex construction entirely subjective in nature. We gravitate toward family, friends, and colleagues, sharing our opinions and absorbing theirs, thereby completing the web of illusion that makes up our daily lives.

One body of opinion that has shaped our experience of reality over the last few hundred years is scientific materialism, which is directly concerned with the perception and measurement of the objective world. Instruments have been developed of greater and greater sensitivity to measure more and more subtle effects. When a new force is perceived and measured, it seems to have philosophical repercussions, which slowly sift down through society, until the fabric of collective consciousness is subtly reconstituted. Perhaps this is due to the vocabulary that invention generates. When Newton's laws of motion were expounded, the vocabulary of push, pull, leverage, attraction, action, and reaction became a way for us to represent reality, and these laws and words spawned a mechanistic view of understanding nature.

While Newton's heritage was a vocabulary of gravity, Einstein's was a vocabulary of light and of a relativity that has profoundly reshaped collective consciousness. Relativity sounded the death knell for scientific materialism, because it made experience of the object dependent on the perception of the subject. Subject and object are a continuum. And just as subject and object are interrelated, so too are body and mind, and matter and energy—with consciousness free to dwell at any point on this duality spectrum. Where before the whole crux of scientific investigation was to be as detached as possible from the object, relativity theory has shown this to be an ineffective and inaccurate means of investigating subtle nonmaterial forces.

This is where astrology comes in as a tool for perceiving reality. Dealing more with the mind and senses of the *subject,* or individual, there is an intrinsic acceptance that the *object*—that individual's experience—is mutually interrelated and interdependent. Rather than life simply happening to us, we are constantly evoking events in a complex dance between our character and our fate, or between our consciousness and the object of our consciousness.

An astrological consultation I gave in the late eighties may serve to illustrate this phenomenon. It was for a middle-aged lady who had a very tenuous grasp of reality, with powerful delusions about being followed by men. I did my best to persuade her that she was probably imagining most of the incidents, based on the astrological fact that she had Pisces rising, with its ruler Neptune exactly on the Descendant, obviously evoking a tendency to be confused in her relations with others. It was an unconvincing consultation undermined by my inability to deal with her mental state. A few minutes after she left the office, I decided to go out shopping, but on opening the door, I found the lady on the stairs studying a bus timetable and muttering to herself. Not wishing to appear to be following her, I smiled weakly and retired to my office, waiting until she had proceeded on her way. Acutely aware that I might confirm her fantasies if I crossed her path, I walked into town by a circuitous route. Twenty minutes later I arrived in the town square, and as I did so the bus pulled up alongside me and my client stepped out. She took one startled look at me and started walking rapidly in the other direction.

Experience had vindicated my client and proved to her that her version of reality was the correct one. The extraordinary thing was that my own behavior had been altered and events had conspired to bring about that which I had wanted to avoid. This scenario plays itself out constantly in all of our lives, as our personal character stamps its impression on a reality that is constantly adjusting to who we are and what we do. The corollary of this is good news in terms of free will. By adapting our behavior, we can alter reality and our experience of it. And everything in our world will alter in it, including the people we relate to. Herein lies the power of astrology, which can be released by judicious work with the energies reflected in the horoscope. And herein lies the possibility of transformation.

The Subjectivity of Perception

Adapting and altering behavior is very, very difficult because it is built on the most basic building blocks of perception. There was an innocent time in childhood, I think, when we could simply see, simply hear, simply feel, simply smell, and simply taste. It did not last long. Our seeing ended when we named what we saw. In using a word, it became a representation for what we saw, and stood in its stead. This is the phenomenon that Magritte called attention to with his painting of a pipe with the famous words, "*Ceci n'est pas un pipe*" (This is not a pipe). The picture of the pipe was *not* the pipe.

When we saw a red-breasted creature flying by the window and were told by a solicitous mother "Bird . . . robin," another nail was hammered into the coffin of perception, because now a robin became grouped into a category together with a crow. It was categorized. We no longer just see the robin, we generalize it unconsciously with its group. As we do with "men," "women," "animals," and everything else. This distorts perception.

The first time we saw an airplane, we were transformed; the second time, we never saw the airplane, because our mind conjured the memory of the first plane in an instant. The first time is always the best. Repetition deadens impression because memory cuts across the senses, although in terms of awareness every time should be equally vibrant. Zen mind—beginner's mind.

Far back in childhood, we lost our ability to see, hear, feel, smell, and taste without this interference of the mind, though we can attain momentary glimpses of the paradise of pure sensory perception. However, if we only dulled perception by labeling, categorizing, and repeating in memory, then we would still have pretty good perception. Perhaps this process is simply the basic expression of Mercury. Nevertheless, a further filter through which we experience the outside world is probably related to the root energy of Venus or perhaps the Moon. When we have an experience of any sort, we are either attracted to or repelled by or indifferent to it. We normally accompany this with an instinctive and unconscious judgment that automatically colors any further experience of the same nature.

The dulling of perception through the nature of our mind, tastes, and instincts is something all humanity shares, and it is a fairly natural process. Matters are complicated, however, by the fact that perception tends to be overlaid with emotion and further clouded by fixed opinions. Our experience of the "objective" world is anchored in the emotional state we were in at the time of the experience. For the astrologer, the predisposition to be in a certain emotional state when having an experience can be related to planetary configurations in the horoscope. For example, a Venus-Saturn combination would see love tempered with duty, or a Mercury-Uranus combination would see traditional learning tempered with restlessness.

Given these predispositions it is almost impossible for an individual to objectively experience what really happens. The individual experiences what he or she *thinks* happens. The basic transformational process in the astrological consultation is to get the client's reality and "real" reality to more or less concur. In other words, the crucial process is to remove the major emotional overlays from the perception of an event so that events are re-experienced more or less how they are generally agreed to have happened, or even make a new interpretation of the event, which is simply more empowering. This is where astrology comes in as a supreme tool, because it is possible to identify perception filters (see figure 1, p. 9), and reveal those that are likely to distort experience in a way that causes pain.

A difficult Mars-Jupiter aspect, for example, may predispose a person to experience dominant males as expressing opinions too forcefully. There will be reasons associated with this predisposition going back to the earliest childhood like beads on a string: a politically conscious father who brooked no dispute; a teacher who browbeat students; a brother who acted intellectually superior. This was the perception of childhood for that person. It may actually have been like that, and it may not. What is important is that the person perceived it as such, and it was vindicated by "reality." Layer upon layer of associations connected with opinions will now predispose that person to be insensitive to or unaware of what is actually happening in almost any discussion later in life. Unbeknownst to this person, to counter the experience of being put down intellectually, precisely those traits that are most feared in others will manifest in the person's own behavior. And it is the privilege of the astrologer to reveal this behavior to the client and inaugurate change.

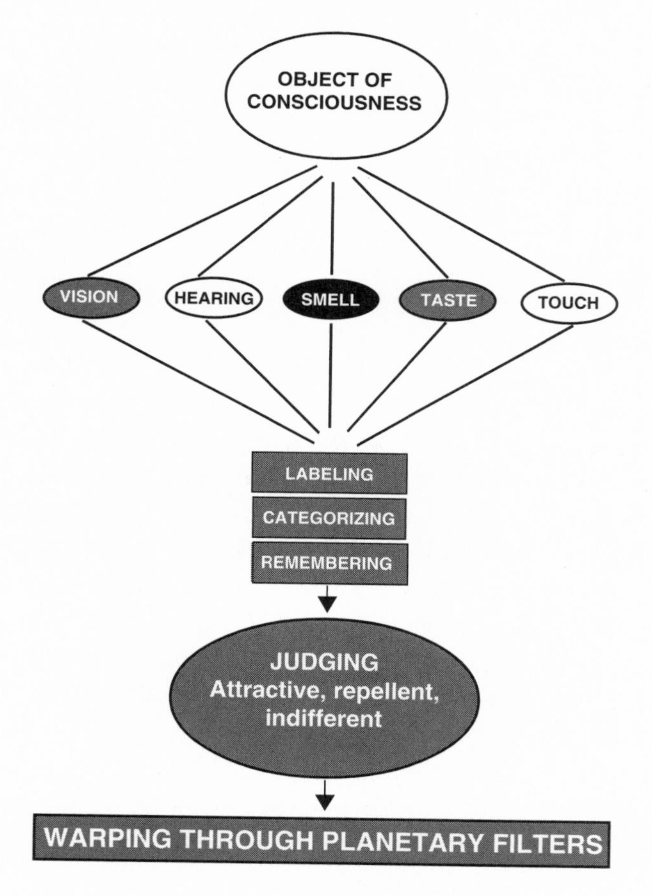

Fig. 1. Filtering reality. Perception of reality happens via the senses. Inherent human mechanisms filter perception, distorting reality. Finally, this filtered reality is further warped through preferences, convictions, and problems reflected in planetary positions in the horoscope.

It should be clear that what the individual experiences as reality is not objective reality. Individually experienced reality did not exist before the individual experienced it and will not exist after—the individual creates his or her own private experience of reality. Given that this is the case, it is amazing that we function as well as we do or that any kind of real communication can take place. But, sharing the same cultural background, schooling, and basic set of beliefs and values, we get by. The sharing of beliefs is so crucial, that powerful emotions are stirred when beliefs are dropped or challenged. Great battles are fought over beliefs, whether it be in the field of science, philosophy, religion, or politics. This is because the more people who agree to share a certain interpretation of reality, the more that reality is validated. This does not make that reality objectively true. But it does make it powerful, and with power comes prestige, success, and economic reward.

The future looks bright for astrology because its philosophical foundations in the understanding of the interrelationship between the individual and the cosmos and its clear bond to something as measurable as the solar system of which we are a part give it the strongest credentials for a workable system. Furthermore, the new understanding of interdependence between subject and object in modern fields of science means that there is something of a rapprochement between two views that have been inimical for a long time.

The Spiral of Time

In astrology, there is a built-in understanding of the dimension of time and its meaning for human development. As everything is in motion, no moment can ever be repeated exactly, though it is eternally paralleled in cyclic motion, creating a spiral evolution through time. The Moon will return to the same degree after a month elapses, but everything else will be in a new position. The nature of the mind is such that we are constantly imagining time as static—as a series of encapsulated events or moments—but there is no instant at which time can be said to stand still, though for comfort we strive to make it so. We tend, then, to define our lives in terms of static concepts, which actually bear little relation to the process of development that is constantly taking place. We use words like "job" to refer to our constantly changing working life or "marriage" to refer to an evolving relationship, and in doing so we create constructs that limit our capacity to act. In fact, as we reconstitute our being each instant, the opportunity for change is present at all times.

In a sense, we recreate ourselves with each imagined moment, basing our identity on who we were the moment before. Again, we are predisposed to recreate ourselves in a certain way, based on our character reflected in the horoscope, but the fact that the universe is in a state of constant flux gives individuals a much stronger hand to create personal transformation through a conscious effort of will. It gives the possibility of making an *intervention*. By following the movement of transiting planets, the astrologer can easily see during which periods particular kinds of therapeutic intervention are most likely to succeed, as there are predispositions for certain kinds of things to happen with different planetary configurations.

Optimal conditions for particular events can be utilized in many fields, from personal development to scientific advances. Nick Kollerstrom's work with the precipitation of metals demonstrated that certain aspect patterns facilitated precipitation.[1] Compounds of iron, silver, and lead, the respective elements of Mars, the Moon, and Saturn, were shown to precipitate strongly with conjunctions of these planets and to a lesser extent with the other major aspects, and this phenomenon was shown to fade out after the aspect had become complete. This illustrates the importance of timing and the value of patience while waiting for optimal conditions. The corollary of this, electional astrology, is an example of the creative use of planetary conditions to achieve desired results at specific times—an example of the proactive use of astrology. Also, using the astrological conditions at the time of the consultation, the best therapeutic strategy will then be to use the current planetary aspects and positions that parallel those of the birth chart.

It would seem that some scientific developments and results can only be attained at certain optimal times, viewed astrologically. For example, on March 23, 1989, the researchers Pons and Fleischmann claimed a world-shattering discovery at the University of Utah when they confidently announced the phenomenon of cold fusion to an astonished world.[2] Instead of the several hundred million degrees normally considered necessary to fuse lighter nuclei into one heavier nucleus, thereby creating vast quantities of energy, these two researchers maintained that they had created fusion at room temperature. Initially this cold fusion process appeared to be successfully replicated in France and in other laboratories around the world; yet, at a subsequent federal conference in May, the process was discredited, and reports of successful replication dwindled. How could reputable scientists make such a mistake?

On March 23, 1989, Saturn and Neptune were in tight conjunction, and this conjunction—which recurs in the heavens once every 36 years—lasted most of the year. Astrologers associate Saturn with form (and cold) and Neptune with dissolution, and, on another level, Saturn relates to experienced fact and Neptune to fantasy and fiction. The conjunction occurred in the sign of Capricorn, which is connected with boundaries in the material world and the drive to overcome them (and, on a more personal level, the ambitious urge to rise in status). As Saturn merges with Neptune an astrological picture arises of the fusion process and the subsequent confusion about cold fusion (and loss of status for our two research scientists!). Was the whole thing fact or fiction? Such is the nature of Neptune, which casts its misty cloak around everything it touches; the truth may never emerge. Perhaps cold fusion really did happen and can only repeat itself during the next conjunction of those representatives of the concrete and sublime, Saturn and Neptune. The important lesson of this event, which shook the scientific world, was concerned with the nature of reality, the nature of illusion, and the difficulty of establishing objective truth.

As Above, So Below

The basic principle of astrology is that the smallest thing in the universe is subject to same process as the largest. The same rules apply for both, and indeed an action in one sphere will reflect an action in the other. That which affects us in our daily life reflects that which affects the universe. Furthermore, time and the physical world are interdependent. Astrology is unique in that it applies rules of correspondence between time and space, linking them to human character and history. Working with the concept of time and character brings an understanding of fate, which in this context is in no way fixed, but interactive with character. If past actions create fate in the present, then so do present actions, putting the individual very much more in control of the time and its material consequences in the world than skeptics might imagine.

With the turn of the millennium, and as we begin to integrate the philosophical consequences of the scientific discoveries of the last century, a transition is occurring. Relativity theory shows the interactivity of subject and object, matter and energy, body and mind. The awareness of unity within duality is arising, and it is in this awareness that a meeting point can be found between the world of rational science and less rational astrology. Astrology cannot be proved satisfactorily using methods of thinking founded on dualistic thought, just as there are aspects of modern science that do not respond well to dualistic logic. Using old-fashioned scientific methods and demanding replication without consideration of the ever-changing cycles of time and its influence on the process can lead both the modern scientist and the astrologer astray. Indeed, the successful practice of astrology is dependent on the awareness of the interactivity between the mind of the astrologer and the object of his or her consciousness.

Astrology ascribes meaning to planetary events and assumes that the energy that moves the universe has a kind of inherent intelligence. The astrologer maintains that there is a natural resonance between the evolving motion of the universe and the development of the human soul. This is a very effective working hypothesis, and the astrologer who puts doubts about its effectiveness aside and embraces the hypothesis wholeheartedly is rewarded by this intelligent universe. The clinical and objective approach of the skeptic will lead to very poor results in the interpretation process, while the enthusiastic believer will become engaged in dialogue with a supportive universe, magically geared to his or her development.

This interactivity between consciousness and a "supportive" universe, and the fact that outer events tend to conform to inner convictions, is at the same time the greatest strength and the greatest weakness of any belief system, including astrology. We create a world around us that reflects our methods for seeing the world.[4] And the world responds intelligently. Events unfold in discrete harmony with the beliefs and conceptions of the observer. Where the scientific approach to the consultation might see the client as the object, the nondual approach sees the sensory and intellectual interaction of the astrologer and client as a unified field, which affects each individual equally. And in this unified field where consciousness focuses its attention on events, meaning arises. Objectifying astrology, and trying to prove it, removes the observer from the very field of consciousness in which astrology works so effectively. Conventional scientific methods may be effective at quantifying the stationary observable universe, but in the mysterious and invisible universe of consciousness—a world that has its parallels in the field of quantum mechanics—the concepts of relativity and paradox come into their own.

Shared Astrological Reality

The truth of astrology will be accepted when the majority of people embrace it as a part of their reality. Today we are very close to this happening, and in all probability future generations will accept the natural correlation between the individual and the cosmos in much the same way as we believe in, say, psychology today. The question

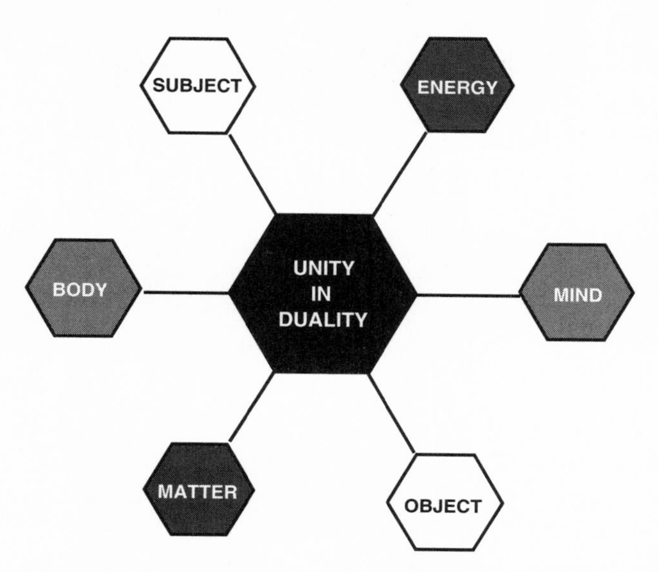

Fig. 2. Unity in duality. Subject/object, body/mind, and matter/energy are polarities that merge when consciousness is focused in the Now, and separate when the rational mind insists on objectivity. When polarities merge, there is unity in duality.

is not whether astrology is objectively true, but whether astrology will become a generally accepted representation of reality. Whatever the case, it should be clear that what the individual experiences as reality is not objective reality. This is why shared beliefs are so important—the greater the number of people who agree to share a certain interpretation of reality, the more that reality is validated. This does not make that reality objectively true. Astrology is simply a very good basis for modeling reality because it is based on the observable planetary system of which we are a part.

Astrologers will be able to produce irrefutable examples of the effectiveness of astrology: obscure predictions that have come true and wonderful correlations from their own life. But their reality is based on the same rules as the client I described earlier. They are being *followed* by the interaction of the world with their own belief systems. It is more relevant to ask whether the system in question enriches their life, whether it harms or does good, than whether it represents truth. The world is vast, and the capacity to extract meaning from it unlimited. Astrology is one way—a very effective way—of doing this. What is more, the astrological view is ecological because it sees humanity and nature as interdependent, and as such it avoids the reductive and destructive consequences that have been the result of the scientific materialistic view.

When we understand that all beliefs are completely subjective in nature, we learn tolerance and respect for the individual whose life is unavoidably circumscribed by personal convictions for which there are inexorable predispositions. If we realize that the sharing of beliefs on a large scale completely determines our experience of reality through a specialized vocabulary, then we will take care not to propound division by introducing strongly held personal concepts, whose truth is merely a reflection of aspects of our character.

Psychological issues of a mental and emotional character cloud our experience of reality. Understanding this, we can commence a clean-up campaign to reinstate the lost innocence of pure sensory experience, which is the closest we can get to a state of purity and bliss, at least in normal consciousness. Reality exists at the meeting point of the subjective mind and its object. Everything happens here. For the individual, empowerment lies in the act of focusing consciousness. When a belief system such as astrology is interpretatively coupled to that consciousness, then knowledge dawns in an interactive process with an intelligent universe. As an astrologer, the degree of your commitment to the interpretative system of astrology will be matched by the success of your interpretation. That is the nature of Mind.

Therapeutic success depends on this commitment. Focusing your attention on any issue with a client, or indeed yourself, will initiate a process that will automatically dredge up the root causes of that issue. In the meeting point of your focus of consciousness and the corresponding reaction of the client—in this charged unified field—change is the order of the day.

Making Contact—Creating the Unified Field with Your Client 1

In this chapter I will describe the basic, initial steps to create the unified field in which you and the client engage in transformation. These steps involve:

- Tuning in to the client and harmonizing body expression;
- Matching the client's voice tone and conceptual expressions;
- Creating an environment in which the client feels comfortable about your competence.

In any therapeutic environment, the relationship between counselor and client is crucial. They each have their own preconceptions, convictions, and emotional reactions, and all these things make true contact extremely difficult. The experience of each party will be channeled through a net of conceptual and emotional filters, all of which are stumbling blocks to true communication and transformation. If these filters are too intrusive, then the synergy is weakened, and the chance for change lost. The advantage for the astrologer is that the horoscope shows these conceptual and emotional filters precisely, and these can be used as conduits so that the consciousness of astrologer and client meet, and it is at this meeting point that the powerful transformative energy can be harnessed. At this point there is no subject, no object, just the subject/object continuum, or put in another way: the relationship.

There are many ways synergy can be enhanced, and many ways it can be impeded. Basically, astrologer and client must be on the same wavelength. This is the job of the astrologer, not the client. With some people it's easy to be on the same wavelength, with others, not. But it is possible to get on the same wavelength with everyone, though not always comfortably. It's important to like and respect clients—it may be impossible to help the person if you don't. If you don't respect or like certain clients, you have not understood them, or they should not be there. It is impossible not to feel for a person with whom you are on the same wavelength.

Subliminal Communication

Studies have shown that most communication is subliminal and that the actual meaning of the words spoken has far less effect than the tone in which words are said. I am sure readers can experiment with saying, "I do love you" and evoke quite different responses. Studies have shown that something less than 10 percent of meaning is perceived by words alone—voice tone and body language are completely dominating.

Body Language

To improve contact there are several levels at which the astrologer, or any counselor, can work. One is the purely physical. People hold their bodies in a myriad of different ways, and there are a myriad of powerful reasons why they do this—some of them connected to their reaction to the person they are meeting, in this case the astrologer. When people get on very well, their bodies move in synch, in an extraordinary dance in which both parties reflect each other's body attitude. Good communication seen in slow-motion video is a finely attuned waltz of body, hands, and facial expressions. The attentive and tuned-in astrologer will automatically assume some of the body language of the client, but there is no harm in helping the client feel at ease by at first consciously registering the client's body language and reflecting it. This is not the same as mimicking, which of course could be offensive; it is rather asking the question, "what does the client feel like that he has assumed this particular pose," and trying it on for size.

In some extreme cases, body language matching is the only way to make contact. After many years of consultations, an astrologer will always get a few individuals who are simply not balanced enough to derive benefit from an astrology session. A person once called me from a pay phone for a consultation. I agreed rather reluctantly, having no means of contacting him if necessary. He did turn up, but it was clear that he was confined within his own mind and simply could not receive. Sitting hunched on the chair, with both elbows on his knees and head bowed looking at the ground, he commenced to transmit. For 45 minutes—the first half of the tape—he talked in a low monotone to the floor. At no time did he look me in the eye. I assumed the position. Short questions uttered to the floor in a low monotone came from me at intervals. This man had had a lonely life. Nobody ever listened to him, and neither did he expect anyone to. For 45 minutes he felt listened to. This in itself was a cathartic process for him. The second half of the tape was a true dialogue—the first he had had in 15 years. It was a breakthrough achieved almost exclusively by matching body language and voice tone.

It is not necessary to match the body language of the client completely, but it is essential not to mismatch it, unless you are deliberately trying to provoke. As I will explain later in greater detail, behavior dominated by air signs and fire signs will express itself in a very erect posture with considerable gesticulation, while earth and water sign behavior is more laid-back, often with the body rather slumped and the eyes looking down. If the astrologer responds to air/fire behavior with earth/water body language, there will be no real communication. In fact, it will not be possible to understand each other. So if the client says: "I just don't see any future in my job" (exasperated voice, *visual* expressions, back straight, both hands forward with palms upraised), and the astrologer says, "That must feel heavy and frustrating for you" (low voice, *physical* expressions, hands folded across stomach, looking down), then that would be a total mismatch. The client would hardly register there had been an interaction. If instead the astrologer sat bolt upright, energetically expressing indignation on behalf of, or understanding for, the client's situation, and used words enhancing the *visual* panorama for the client: "So do you see anything better on the horizon?" then astrologer and client would click.

Voice Tone

It is often possible to guess at voice tone just by looking at Mercury in the horoscope. In a fire sign and aligned with a planet like Mars or Uranus, the tone will be fast and staccato and slightly higher in pitch. This person will search the air with his eyes, as if dragging ideas from the ether, and will be full of sudden movement and gesticulation. On the other hand, a Mercury conjunct Saturn in Scorpio will denote a person of few words, which when uttered will come out painfully slow. This person will search within to find words and will carefully censor what can be said in a slow, thoughtful process. The voice will be deep, the tempo slow. If Pluto is configured more strongly with Mercury, then these give-away signs will, however, not be present. The eyes will either rivet you in an intense stare, or look at some indefinable point without flickering. The voice could be more monotone, rather deadening, and perhaps extremely tiring to listen to for any length of time. Configured with Venus, the voice will be melodious and a pleasure to hear, perhaps rather seductive, very much feeling its way in a constant checking that the communication is going down well with the other person.

As with body language, it is not essential and perhaps not possible to match voice tone exactly. But it is essential not to mismatch. Someone with Mercury in Gemini sextile Mars will lose patience with an astrologer who embarks on slow, laborious explanations. They would respond much better to a fast ping-pong. Conversely, a fast and breathless discourse on the dynamics of some astrological influence would cut no ice with a person with a thoughtful Mercury-Saturn conjunction in Cancer; this person needs to have the emotions touched first, before understanding dawns.

Targeted Communication

It is essential, then, that you pitch your communication at the client's tone and conceptual level. You can do this reactively or proactively. Reactively is to note the tempo, tone, and content of the communication and to respond on the same level. Proactively is to choose a tempo, tone, and content precisely attuned to the astrological influence that you wish to address. Proactive communication of this nature is an extremely powerful tool in the astrologer's arsenal. Finely adjusted questions using the words and mood of any particular planetary configuration will immediately strike home in the consciousness of the client, saving valuable time in the consultation process. For example, if the astrologer wanted to home in on a client's Mars in Leo conjunct Pluto, the astrologer would immediately strike a chord with, "It must be quite a jungle out there for you—I imagine that takes considerable survival skills, right?" You have entered the client's conceptual world or metaphor, and your words have touched Mars in Leo's need for respect; this client will now be dying to open up about quite a difficult area.

Being Believable

First-time clients are going to be rather nervous, and, above all, they are going to be somewhat skeptical. They are not sure that astrology "works," or if they are sure, they may have unrealistic expectations.

Men in particular tend to assume a defensive posture, especially if emotional issues are touched on. As a general rule, men want to talk about their jobs, and women, their relationships. Ideally, both sexes ought to be equally interested in both areas, but experience shows they are not. It is not appropriate to work with psychological transformation on someone who does not express an interest in it, and the astrologer should be prepared to make a competent consultation without therapeutic intervention. Having said that, everyone limits their potential to live life to the fullest because of personality traits reflected by astrological configurations. Negative traits and behavior will always have unwanted consequences in a person's life, whether it's in relationships or in business. If you can help the client identify these consequences, then it is not difficult to get the client's agreement that they should and can do something about them. Then it is a short step for the client to actually express a desire to work psychologically on a particular theme.

It is crucial at the very beginning of a consultation to win the confidence of the client. Nobody wants to fully reveal intimate behavior patterns to a person they have just met, unless that person gives a deep impression of competence. Astrologers work in different ways, and some prefer a consultation that stretches over two sessions, whereas others even have a course of sessions stretching over several weeks. In these cases, intimacy can be built up at leisure. Most astrologers I know, however, see a client in one longish session, then don't expect to see them again for some months or perhaps a year. This means that they have from one and a half to two hours to have an impact, and the nature of astrology is that it is possible to do this if the client is reasonably well-balanced.

For my part, I spend at least the first 10 to 20 minutes of a consultation using pure astrological techniques to identify themes in the client's life.[5] Combining the horary chart for the client's arrival with conclusions gleaned from studying transits and progressions, it's possible to accurately identify live issues in the client's present life, trace their beginnings, and surmise their endings, without any feedback from the client. I tend not to even look at the client at this stage. This is partly through the desire to establish the credentials of astrology alone and partly because I don't want to be seen to be interpreting subliminal signals that could subsequently be supposed to have guided me in my initial description of the client's situation. This makes subsequent therapeutic work much easier because, from the client's point of view, if deep issues can be identified without the client saying anything, then the client sees no reason to impose constraints on subsequent communication. The client thinks the astrologer must know anyway.

This initial period is crucial. The client must have confidence in the skills of the astrologer. We all tend to deny our less palatable behavior, even though we know that we can only progress by bringing it out in the open. This behavior will remain concealed unless the astrologer has asserted his authority and inspired trust. Some clients are very forgiving, other are not. Often a person with strong Scorpio traits, for example, will completely lose respect for an astrologer who pussyfoots around. Their secret agenda is to come through the consultation without revealing a thing—a quite exhausting process for the astrologer. This type is actually extremely grateful when strongly confronted by definitive statements—indeed, you will get nowhere if you don't.

I remember phoning one client prior to the consultation, who had a Mars-Mercury-Pluto conjunction in the 6th house (see figure 3 on p. 24). I was a little unsure whether I had the correct birth information.

When she arrived, she made a remark that I was *checking up* on her to see if she really was coming. When I said that was not the case at all, she implied in an off-handed way that I wasn't telling the truth, but that was OK with her. She expected people to lie! This is exactly the kind of suspicion that is fuelled by a Mars-Mercury-Pluto conjunction, evoking a level of paranoia and a conviction that information is being hidden. It is not difficult to imagine the kind of effect this has on her working life and collegial relationships, considering the 6th house influence. Rather than politely continue, I chose to confront her on this issue and spelled out in detail the behavioral flaw and its consequences in her working environment. Recognizing these consequences, which only make her unhappy, she was able to understand her unconscious need to make other people ill at ease in an effort to establish control. We were five minutes into the consultation and had already gotten into the core issue.

Opening remarks from the client are invariably very significant, like pearls handed to the astrologer to place on the scales. Does the client make some kind of excuse on arriving, tell of some travel difficulty, try to make *you* feel at ease, stride into the room and sit down (in *your* chair), or nervously await instructions? All these things reveal much about the character. But the opening remarks have special significance, though confronting a client about them in the initial phase would be the exception, rather than the rule.

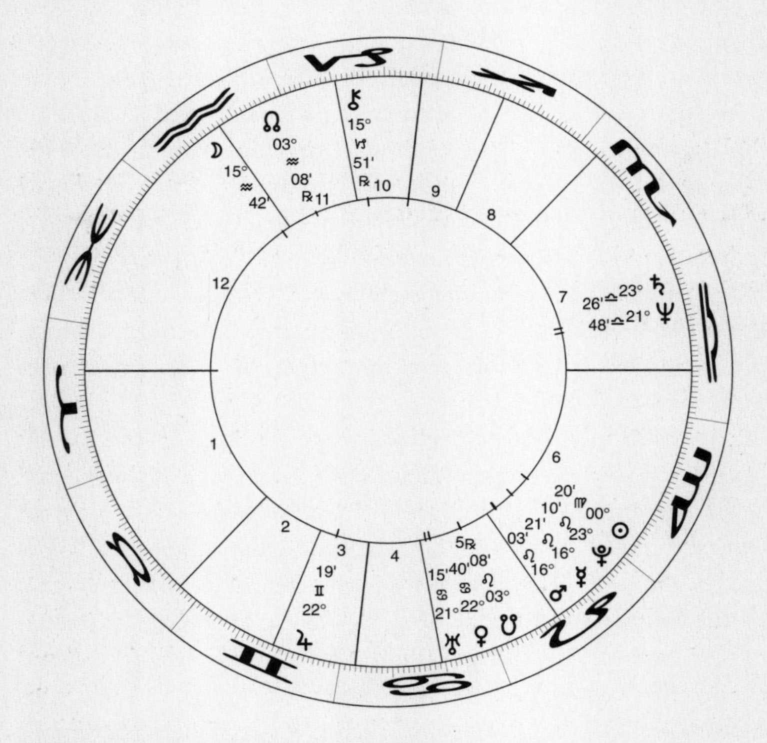

Fig. 3. Gitte M. A client with overforceful communication.

2 Communication—Getting to Your Client's Core Issues

During a seven-year period in which I ran a school for training astrologers in the consultation process, I sat in on many trial consultations. Students were encouraged to engage in a dialogue that was mutually enriching rather than make long monologues about the supposed influence of planets, aspects, signs, and houses in the life of the client. Ultimately, the one-way consultation is an exhausting process for both astrologer and client. The astrologer can never learn anything new, and the client can only nod in agreement (if the astrologer happens to be correct). What is most interesting in a consultation is what *clients* have to say, because they furnish the astrologer with the material to go into depth in any particular issue.

In these trial sessions, students often ground to a halt after following a particular avenue of investigation for a minute or two. They baited the hook, the client took a bite, but then escaped. It was impossible to get the client to bite again in that particular area. Missing an opportunity like this should be avoided. When addressing any astrological configuration, the practitioner should pursue its ramifications until the core issues eventually emerge, rather than leave the issue hanging unresolved. This requires specific communication techniques, which have to be mastered. This chapter will go through some simple rules of communication and show the intervention techniques that can be used to get to the core of any astrological issue within two to four minutes. These communication techniques have been pioneered by the early practitioners of NLP (neuro-linguistic programming).[6]

Dialogue and Intervention

One trial consultation with a female astrologer in my school went like this. She had a client who had a Mars-Neptune conjunction in Scorpio in the 6th house (we'll call her Mary):

Astrologer: There is an influence here that shows disappointment in your working life, perhaps connected with men.

Mary: That's true. You can't trust men at work.

Astrologer: Oh? Why not?

Mary: Well, they're only after one thing, right?

Astrologer: Ah! [Laughs] Well that could be a problem.

Here the astrologer has made two basic mistakes that completely derail the effort to get to the bottom of what Mars-Neptune in the 6th really symbolizes. The first mistake was asking the client to explain *why*, and the second was getting *enrolled* in the client's view.

In fact, it is rarely necessary to listen to more than a few statements from a client before a communication intervention can be made. Client monologues that last over five minutes serve very little function whatsoever, apart from making the client feel cozy (not always a good thing), and providing (or perhaps inundating) you with background information. There is certainly enough material in the above interchange to vigorously proceed to the core issue. The secret of precision communication is based on one simple thing, and that is that the astrologer must ask: "Is there anything that this person has just said that I do not understand or is not explained 100 percent?"

The statement, "You can't trust men at work" has two, perhaps three, areas not fully explained. In particular, we don't know who "you" is, and we don't know who "men" refers to. This can be sorted out with the following questions: "When you say 'You,' Mary, do you mean yourself personally?" After all, it's safe to assume here that some people would not agree, and would allow that some men, somewhere in the world, can be trusted at work. Mary would have to *own* the statement: "*I* can't trust men at work," and in doing so, a small chink in her conceptual armor would appear.

"Men" is a nominalization that groups half the world's population under one umbrella. "Men" conceals more than it reveals. There are several ways of approaching this, and one would be: "Mary, do you know of any men anywhere who *can* be trusted at work?" Mary would have to agree that some men somewhere could probably be trusted, and another chink in her conceptual armor would appear. Some men can be trusted, some men can't.

With the third statement ("Men are only after one thing, right?"), the client tries to enroll the female astrologer in her views about men, and the astrologer is lost if she allows this. There are two things we are not 100 percent sure of: (1) who these men are, and (2) what they are after. Let's check: "So what exactly are men after?" OK, Mary will raise her eyebrows at your naiveté, but let's hear it: "*Sex*, of course." (Well, it could be power, sympathy, money; don't get enrolled in the client's assumption.) Then the next question: "Which men in particular are interested in sex?" Now the answer might be: "Well, it's Rupert in the office; he's always making sexual innuendoes."

What is important here is that you now have a specific statement describing a situation that can be dealt with, rather than a generalized statement that circumscribes and limits the client's options and locks her up in a conceptual prison. This process of going toward the specific and away from the general is the crucial process in getting to core issues. It might seem as if trivial questions are being asked here, but this is not the case. These questions are like a scalpel that cuts through surface communication and gets to the very heart of the matter. It is a very powerful tool and should be used sparingly. Once you have arrived at the core, this tool must be put aside and other techniques employed. Having come through to the internal organs, as it were, it's important to start the work of repair. But initially, in-depth penetration is necessary, and words are the way to do this. The communication techniques described in this chapter are powerful, and they can be perceived as a form of interrogation, so there is no need for this process to go on for more than a few minutes, otherwise they may arouse antagonism. Most deep issues can be brought to the surface within this time, and the beauty of it is that the client supplies you with exactly the material you need.

In asking these questions, it's wise to be concerned with the *how* rather than the *why*. Asked to explain why, a client can ramble on forever, muddying the issue. When asked how, the client must actually reveal the structure of inner identity. So the kind of questions to ask are: "How precisely . . . ?"; "What prevents you from . . . ?"; "Can you expand on . . . ?"; or, when you sense that the client leaves a sentence hanging: " . . . but? . . ."

Cutting through Language Barriers

Core issues are concealed by surface communication using a broad variety of grammatical forms, which can be categorized in specific ways. It may seem tedious to learn them at first, but having grasped these grammatical forms you will never lose the fish from the hook, and with a few skilled twists of the rod, you will land the creature that reveals what really is going on beneath the surface. The point is that every grammatical form invites a specific kind of question and relates to specific astrological patterns. When you can identify the nature of the statement (for example that "Men" is a *nominaliza-*

tion), you can locate the astrological pattern, and guided by the astrology, all the ramifications of a simple statement can be pursued. The following section describes the main grammatical forms and language structures that you should watch out for.

Generalizations

Socialists are only interested in penalizing the individual.

A person with a Sun-Pluto conjunction in Leo in the 11th once said this to me. Political generalizations of this sort are often related to 9th-, 10th-, and 11th-house issues—especially the 11th house. The essential idea in dealing with generalizations is that it does not matter whether the client is right or wrong or whether you agree with him. The only issue of relevance is what is it at the core of the client that has inclined him to perceive things in this way. Some inner compulsion has over many years formed into a viewpoint, which is now expressed as a generalization, and this generalization impedes the client from a full experience of reality. A generalization is a conceptual filter, that isolates a person by limiting communication options.

The advantage for the astrologer in the following interchange is that it is recognized that the conceptual filter is connected with Sun-Pluto in Leo in the person's 11th house. The core issue *must* be connected with anxiety about self-assertion, however much the client may try to disguise it as a political viewpoint. Bearing in mind that this statement contains language structures that carry meaning the astrologer cannot understand 100 percent, there are a number of questions that could and should be asked: Who are socialists? Who are individuals? Both terms are nominalizations, but we just want to deal with the generalization here:

Astrologer : Would you be thinking about the *present* government? [Specifying]

Client: Yes, them too.

Astrologer: And they have penalized *you* in some way? [Specifying]

Client: Indeed, they have.

Astrologer: Actually the present government gave *a friend of mine* a grant to start her personal business—so they don't penalize *all* individuals do they?"

Client: Well, OK, obviously not all.

It may seem small, but it's the beginning of an important excursion into the convoluted dynamics of the Sun and Pluto. Bear in mind that this type of person will have incredibly strong opinions that are unshakable, because to drop them is to lose a pillar of identity. This will have social consequences because the man will be very dominating among friends and acquaintances. To be his friend, you will *have* to agree with him, and this means that the man will not know who his true friends are.

There are many examples of generalizations, and the best way to deal with them is to get the client to see that there are exceptions. Generalizations are *never* as true as they sound, and they *always* limit perception. Here might be a typical statement from a man with Venus square Saturn:

Client: Women don't like me.

Astrologer: Can you give me an example of *any* woman that has liked you?

Client: Well, obviously my mother likes me.

Basically, this will be a low-self-worth issue projected onto women. Indeed, in all likelihood the sentence could be effectively reversed: "But you like women . . . ?" (see "Inversion" below).

Or a person with Sun square Saturn:

Client: I never receive praise.

Astrologer [sneaky]: Well, I must say from what you've

told me about your job, you're obviously a very hard-working and conscientious person.

Client: But I never get anywhere.

Astrologer: Did you notice I was praising your good qualities just now?

Client: Well, I suppose you were.

Astrologer: So could it be true that you do get praise but don't notice it?

The point here is that Sun-Saturn people tend to have difficulty receiving or accepting praise.

Inversion

My husband is always arguing with me.

Sometimes the structure of a sentence is such that it implies its opposite. Whenever there is an *interaction* between two people, what is true for one of them may well be true for the other. For example, it takes two people to argue, or to emotionally interrelate in any way. It is very powerful to turn the sentence around when appropriate:

Astrologer: Do you also argue with your husband, or what?

Client: Well, of course, I have to respond!

Now, if Mercury is in Libra sextile Venus, then there is probably nothing to this, but if Mercury is conjunct Mars and square Jupiter, then you know that the client is at least 50 percent responsible for the tension.

There are many variations of this theme of inversion. For example, someone who has Sun-Mercury in Pisces in the 8th, perhaps (feeling invisible):

Client: My boss never talks to me.

Astrologer: But you do talk to your boss, don't you?

Or perhaps someone with a heavily aspected Moon:

Client: My partner just doesn't show any emotion.

Astrologer: And what emotions do you show him?

As long as the client thinks someone else is at fault, there will never be a change in the situation. No change can take place without the client taking responsibility. I say to the client that they are at least 50 percent responsible for any given situation. (Actually, they are 100 percent, but there is no need to daunt them!) By changing the quality of the energy invested in any planetary configuration, both the client and his or her environment will change.

Coercion

I have to stick out my job to the end.

Some people have a strong tendency to use words that imply necessity or coercion (technically, these words are called modal operators). Expressions such as "must," "have to," and "ought to" imply that the person feels that something bad will happen if they do not do something. In this case, the language technique you use is to ask what exactly it is that will happen. Often it's in the mind-set of the person, and when he or she examines the situation, the person realizes either that unconscious fear is the driving force or that there is, in fact, nothing to fear. Typically, astrological patterns featuring Saturn (duty) or Pluto (anxiety) lie behind these mind-sets. It is important to pick up on implications of coercion and clear them as they arise:

Astrologer: Who says you have to stick it out to the end? [Sounds like the client picked up this conviction from an early authority figure—you might be looking at a Mars-Saturn aspect here.]

Client: I just feel I have to. [Doesn't bite]

Astrologer: So what exactly would happen if you didn't stick it out to the end?

Here the client is forced to think the unthinkable. There is a likelihood, driven by the dogged conviction he has expressed, that the client simply never knows when to stop and that this taints all his life experience. It would be important in this instance to follow through, so:

Client: It doesn't bear thinking about.

Astrologer: What doesn't bear thinking about?

Client: What would happen.

Astrologer: What *would* happen? [Don't give up! The unexpressed has the capacity to evoke more fear than the expressed.]

It's always worth picking up on sentences that imply coercion or necessity. For example, the statement, "I can't miss the news," might be quite innocent. However, if this person had Mercury tied up with Uranus and Jupiter in a mutable T-square, then there is a strong likelihood that the statement implies a need for constant distraction that could inhibit the person's ability to be calm. Or for someone saying "You *have* to listen to what others have to say," there might be a much deeper significance if there was a stellium in the 7th house involving Mercury and Saturn. The words "have to" suggest that there is a behavioral stricture imposed from the outside, and the astrologer is well advised to find out where it comes from. The question to ask in these cases is: What would happen if you did (or didn't) . . . ?

Depersonalization

One has to be on one's best behavior at family gatherings.

The use of words such as "one," "you," or "people" is a way of avoiding personal responsibility for a behavior or a viewpoint. It is often the result of a prominent Saturn, which indicates that certain social norms dominate individual needs. The crucial intervention here is simply to get the person to replace the impersonal word with the personal pronoun, so:

Astrologer: Do *you* have to be on your best behavior then?

Client: Yes, me too.

But it's not enough. The person has to feel what it's like to actually say the words, so:

Astrologer: So can you say the same sentence using the word "I" instead of "One"?

Client: OK. *I* have to be at my best behavior at family gatherings!

Of course, if you are looking at Saturn in the 4th house, you know there is more to the issue than meets the eye, and this will give you the opportunity to delve into the whole question of family atmosphere and duty, which is, after all, Saturn's favorite way of creating obedience when placed in the 4th. The consequences of family discipline will obviously have repercussions stretching into the client's present emotional life. The obvious question would then be: What would happen if you weren't on your best behavior?

Comparisons

It's much better to make a clean break and get a divorce.

The use of comparative adjectives like "better," "worse," and "more" always implies a yardstick with which something is being compared. You do not know what this yardstick is unless you ask. The point of doing this is both to find out what really is going on under the surface thinking and to check out what value systems are important for the client. In this case, "better" seems to refer to a memory of an experience that must be avoided, and it pays to find out what it is.

Perhaps this client has Moon-Mars in the 7th in aspect to Neptune and saw her mother's marriage as one of suffering. So:

Astrologer: It's better than what?

Client: Better than making a commitment and then being let down.

Of course, you know that this client will never be able to have a relationship without confronting and solving the issues of martyrdom and deceit reflected by the Moon-Mars-Neptune, so there is work to be done showing that the "clean break" idea may just be an escape route. Thus the trick with comparisons is to identify the yardstick.

Enrollment

Obviously, my father was just lazy.

Clients will use many language structures that ask you to take something as given. The whole edifice of the client's reasoning will crumble if you don't, and it's tempting to go along with and humor the client. However, if you accept the statement without question, you can miss an important intervention. In this case, what is obvious to the client is not at all obvious to anyone who does not know the client's father. If you are looking at a person who has Sun in Taurus sextile Mars in Pisces, it is possible that he experienced the father as lazy. On the other hand, the father may have known how to appreciate and enjoy life. Anyway:

Astrologer: Obviously for who, exactly?

Client: Well, it was obvious to me.

There is a crucial difference, because here the client is recognizing that it's his own subjective perception and not some kind of generally accepted truth—and that's something you can work with.

These enrollment attempts occur in many guises. For example, a person with a Sun-Jupiter opposition could say, "Naturally, the school board got it wrong again." The client wants you to go along with his story, but you know that—with his strong Jupiter—he always likes to think that he gets it right. The client is used to other people agreeing with him, hence the word "naturally." The temptation is to say, "Ah, yes," just to keep him happy, but:

Astrologer: In what way was this natural, then?

Client: Well they always get it wrong, don't they? [Generalizing]

So you've got him on the generalization, and it's a small step to get him to admit that the majority of people, having elected the board, think the board gets it right at least some of the time. Don't get enrolled.

Impossibility

It's impossible to have a relationship with someone who's unfaithful.

When clients use words or phrases that imply that they cannot countenance something, they are often concealing deeper layers of anxiety. These kinds of statements can either be related to strong principles, and hence a strong Jupiter or Saturn, or to something that the client is terrified of, reflected perhaps by Pluto. An inability to countenance something bears witness to an abyss in the psyche. If the above statement was made by a man with a Venus-Pluto conjunction in Leo, then the astrologer would know that pride and fear were the real reasons behind the statement. First of all, let's get the client to own the statement:

Astrologer: Impossible for who?

Client: Well, I would find it impossible.

Then the important question:

Astrologer: *What specifically prevents you* from having a relationship with someone who's unfaithful?

Client: I couldn't respect someone who was unfaithful.

Astrologer: Couldn't respect someone who went off and had sex with someone else? [Get him to specify "unfaithful."]

Client: Right!

Astrologer: What prevents you from respecting them?

Client: Well, she's not showing any respect for me, is she?

This opens up the opportunity to find out who that "she" is and to examine what—for Venus-Pluto in Leo—"respect" (a nominalization) really means. When the word "respect" is used in a Pluto context, contempt is often seen on the flip side.

It's crucial to show someone who claims that something is impossible that it is in fact possible. Whole areas of experience are cut off when something is thought impossible; indeed, it is often only by embracing the impossible that healing begins. Perhaps nothing would be better for the above person than forgiving the person who was unfaithful, especially as with a bit of probing it's conceivable that our Venus in Leo person has succumbed to the occasional romance himself. The key question is : "*What prevents you . . . ?*"

Nominalization

I have great respect for women.

A nominalization is any noun that has no physical reality. In this sentence, "respect" is a nominalization—it refers to something unspecified. Politicians love using nominalizations because the interviewer can never tie them down. Once you accept the nominalization, you completely lose your bearings. Consider the statement: "Financial considerations in the ministry are a responsibility factor in the continuing struggle for income parity." There are at least five nominalizations here, and it is really boring to listen to. Good interviewers are trained to crack open this kind of language, which in this case means: "There's no money, and we are responsible, and though we keep trying we can't insure equal pay."

In the astrological consultation, it's safe to assume that behind a nominalization lies a specific thing and a specific event. The person who said to me "I have great respect for women" was studying gynecology, so it was pretty important that he *did* respect women.

But "respect" is a loaded word, especially in this case, where there is a Venus-Pluto conjunction in Virgo in the 5th house in opposition to the Moon and Saturn in Pisces (see figure 4 on p. 39). No *astrological* evidence of true respect here, so:

Astrologer: Which women in particular do you have great respect for? [Asking for a specifying of the generalization "women"]

Client: Well, I really respect my mother. [Read "fear" here . . . Moon-Pluto]

Astrologer: What precisely do you respect about your mother?

Client: She managed to run the family very smoothly after my father left.

Astrologer: So you didn't give her any trouble then?

Client: Not very often. [Implies "Yes, sometimes"]

Astrologer: So what did she do when you misbehaved?

Client: She was just disappointed, that's all. [He's mind-reading here: we have to know what actually happened.]

Astrologer: How did that manifest itself?

Client: She seemed upset and ignored me.

With Pluto aspects to the personal planets, freezing people out is a favorite manipulation technique, and people will often mention

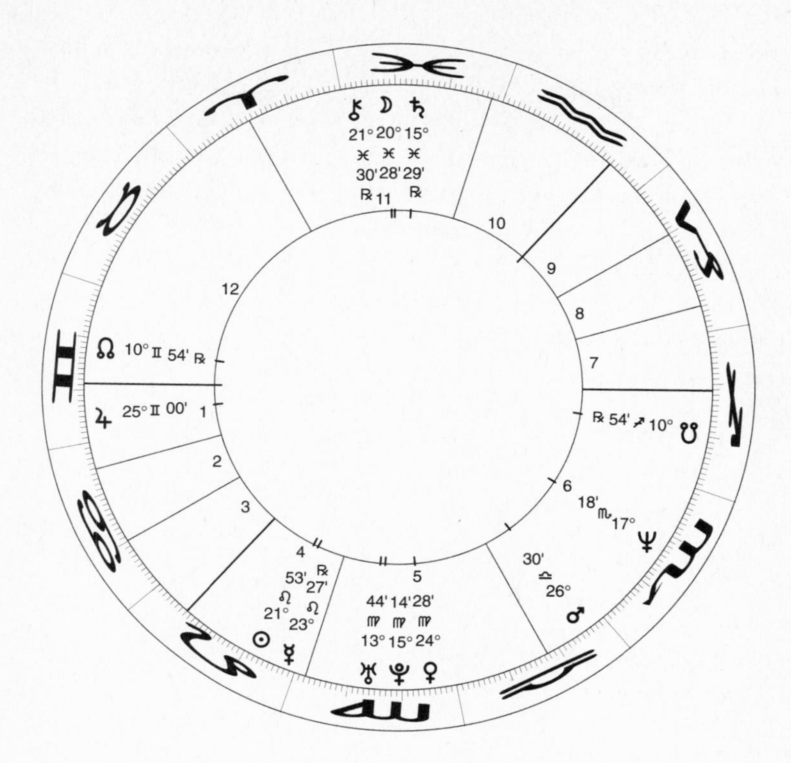

Fig. 4. Jesper D. Respect for women is revealed to be connected with anxiety.

---

being ignored. But what you understand as being ignored and what they understand as being ignored are often miles apart. You have to find out *how long* the freeze-out lasted. The answer is often surprising:

Astrologer: For what length of time could she ignore you?

Client: Well, maybe a few days . . . a week, perhaps.

Now that's an interesting way to run the family! Because this was the atmosphere of his childhood, he does not know there is anything amiss, and it is not until the astrologer demonstrates the consequences for personal relationships in the client's present emotional life that he'll truly understand and want to work with the old energy. Actually, this person subsequently gave up gynecology, which may well have saved a number of women from unnecessary surgery.

With the horoscope in front of you, you could tell the client right at the beginning, "Well, this aspect pattern suggests that you don't actually respect women but, on the contrary, feel a lot of anxiety in their presence." However, this kind of delivery of the interpretation is ineffective as therapy because it will evoke resistance. When the story is drawn out in the client's own words, then the truth dawns for them. So, although the horoscope gives the opportunity to jump ahead, it is really important to guide the client step by step, drawing inferences more from the client's responses than your knowledge of what the horoscope shows. Resist the temptation to jump ahead and second-guess your client.

Nominalizations generally cover an ongoing series of events rather than a simple fact. For example, when someone says, "My marriage is rather difficult," "Marriage" is a nominalization covering the whole process of relating. Marriage is meaningless in this context, so the question would be: What is it about the process of relating with your wife that is difficult? Now you will get facts that you can work on. Another example could be, "My decision to leave my job is irrevocable." "Decision" here represents a reasoning process that at some point solidified, so the question would be: What was your reasoning behind that decision?

Another example could be, "My son's beliefs really bother me," where unspecified "beliefs" is meaningless, so you would ask, "What is it that he believes in that bothers you so much?" Or, "The tension at work worries me" could be opened up with "What is it precisely at your work that seems tense?" You have to bring the client back in touch with the actual sequence of events that is taking place; only then can progress be made. You cannot relate to an unspecified noun, but you can relate to a process.

Exaggeration

You can never trust men.

The client's use of words like "never," "ever," and "always" often shows a decision she made that is guaranteed to limit her experience of reality. It's a denial of possibility, an act of pulling up a mental drawbridge against a perceived fear or threat. It's typical Pluto language, presenting issues as either black or white. It implies either that the client has had a series of experiences that have culminated in a limiting decision or that she has swallowed something raw from an influential person in her life. It is simple to show the emptiness of the assertion:

> **Astrologer** (Male): Do you feel you can trust me? [To regular client]
>
> **Client:** Well, I can trust you, of course.

And thus the statement is proved false. Of course, you will want to find out who "men" (generalization) refers to specifically, so:

> **Astrologer:** So you can trust some men, right? [Client nods.] Who, specifically, can you not trust?

I remember a consultation with an attractive young woman who happened to earn her income through prostitution, recruiting female students (who needed to pay for their studies) into this timeless profession. She did not have a very high opinion of men's ability to resist the charms of women. She had a Mars-Pluto conjunction in Virgo in the 6th house, confirming that this was obviously a service industry. Anyway she didn't trust men, not even me, and brought me the delightful statement:

> **Client:** Men will *always* go to a prostitute, given the chance.
>
> **Astrologer:** Do you mean *all* men?
>
> **Client:** Yes, that's my experience.

Astrologer: So there's 5 million people in Denmark [where the consultation took place], and about a million and a half sexually active men, right?

Client: I suppose so.

Astrologer: Now, in Denmark, the chance is there all the time, right? [It's legal.]

Client: That's true.

Astrologer: So are you saying that 100 percent of all these men will sometime in their life visit a prostitute?

Client: Well, not 100 percent of them.

Astrologer: Well how many of them?

Client: Well, maybe 25 percent.

Astrologer: So over one million men in Denmark will *not* go to a prostitute during the course of their life, right?

The point is this woman had structured her life so that the only men she has anything to do with actually go to a prostitute. This is the power of the Mars-Pluto filter in her life. She has chosen a profession in which her relations with men breed contempt, but this says more about the formative experiences in her life than about how men really are.

You can best deal with definitive words like "never," "ever," "all," "any," and "always," by reflecting the exaggeration. So if someone says, "Nobody ever listens to what I say," you can respond, "Nobody ever, ever, ever listens to you?" Or if someone says, "I never have any success at anything," you can respond, "At no time in your life have you ever had any success with anything at all?!" When people finally agree that their exaggeration is not a reflection of the truth, new avenues of possibility open for them, which can be nurtured and expanded. If it dawns for the person working as a prostitute that men truly exist who can be trusted and that it is her own choice to relate to men who can't, then she can choose an alternative course of action to have a satisfying relationship with men in her personal life.

Cause/Effect

My wife's demands are driving me crazy.

This type of language structure implies that what another person does causes a specific state of mind in the client. This is rarely true, though it may seem so. What actually happens is: (a) Somebody does something; (b) The client chooses to react in a specific emotional way, based on the client's own predisposition. As long as the client fails to understand that he can choose his reaction or course of action, then he will feel powerless. It's a desperate situation for people when they feel their actions depend on another, and it can be a source of empowerment and relief when this is demonstrated not to be the case. So:

Astrologer: What demands exactly? ["Demands" is a nominalization.]

Client: She expects me to visit her family every weekend.

Astrologer: Then you go crazy? How does that manifest?

Client: I get angry and depressed, and we argue.

Astrologer: So when your wife wants you to visit her family, you get angry?

Client: Yes.

Astrologer: Is she *making* you angry, or is that just the way you happen to react to her demand?

Now if this person had, say, Moon in Aries in the 7th, square Saturn, we could understand that he instinctively reacts this way, but he's going to have to work on this if he wants a happy marriage. One could almost say his wife is giving him the opportunity to work on and transform an unwanted pattern of behavior.

These cause/effect assumptions are very common. Another example could be, "My mother's criticism is so irritating." The response could be:

Astrologer: She criticizes you, and that irritates you? [Personalizing what was unsaid]

Client: Right, that's what I meant.

Astrologer: How does she criticize you specifically? ["Criticism" is a nominalization.]

Client: She says I don't spend enough time visiting her.

Astrologer: So she says you should visit her, and you get angry? [Separating supposed cause from supposed effect]

Now if this person happened to have a Mercury square Moon conjunct Mars, we would know that he tends to respond to conversation with his mother with irritation. Then the idea would be to show how there are other interpretations of his mother's behavior (she loves him, for example), and other ways to respond.

False Causation

I'm so depressed my wife doesn't love me.

This type of language structure surfaces when the client says two different things in the same sentence suggesting there is a relationship between them. He is saying here that there is equivalence between his feeling depressed and his wife not loving him. Maybe this is plausible, maybe not, and this is the way to find out:

Astrologer: How, specifically, do you know your wife doesn't love you?

Client: She never kisses me when I return home from work, for example.

Astrologer: So her not kissing you makes you feel depressed? [Drawing attention to the equivalence]

Client: Yes, that's right.

There are now several options. You could try reversing the equivalence:

Astrologer: And if *you* didn't kiss *her* when she came home, would that necessarily imply you did not love her?

Or you could try to find out what the process is exactly:

Astrologer: How, specifically does her not kissing you make you feel depressed?

Or you could make a challenge:

Astrologer: So if she did kiss you, you wouldn't feel depressed any more?

If you are dealing with a person who has, say, a Saturn-afflicted personal planet, very little needs to take place before the client reads his partner's actions as lacking love. Unfortunately, this way of filtering experience ends up being self-fulfilling, so that the client has an abundance of "evidence" about his unloving wife.

Excuses/But

I would move out, but my mother is ill.

This language structure—technically called *complex equivalence*—implies that the person would pursue a particular course of action if it were not for a set of circumstances. Sometimes it's believable. In this instance, one could understand the need to care for a sick mother. But if the client had a Moon-Neptune conjunction in the 4th house we'd want to check:

Astrologer: So how long has your mother been ill?

Client: Well, she's been feeling poorly for many years.

Astrologer: What, specifically, is wrong with her?

Client: She gets migraines and suffers from depression.

Astrologer: But as soon as she's not depressed any more, you'll move out, right?

The point with this last statement is to challenge the client with the alternative. The likelihood is that even if the circumstances changed, she still would not move. Obviously this sounds like a symbiotic relationship of mutual dependence, which is not the ideal course of development for an individual. At this point it would be comparatively simple to illustrate that the client is using her "sick" mother as an excuse to remain at home, but the first step with complex equivalence is always to present the client with the challenge of what they would do if the circumstances changed.

Another example could be, "I'd love to study, but we have to have two incomes to keep the family running." The need for extra income is used as an excuse for not studying. There would be several ways of dealing with this:

Astrologer: It's impossible for the family to survive on one income when you study?

Client: Right.

Astrologer: So you don't know any families who survive where one partner studies?

And of course she does and will have to admit that the possibility is definitely there; it's the priorities that count. What she actually means here is that material needs have priority over intellectual needs, and she either accepts that as true or takes the necessary steps to change. If she really wanted to make the change, you could proceed by asking, "Well, what *exactly* would happen if you only had one income?" and help her examine her fears, one by one.

Mind Reading

> *The others feel I take too much time up in the group.*

Mind reading leads to confusion and alienation in the long term. It's often a dominant language structure when Neptune is strong in the horoscope. Clients claim to have knowledge of what is going on in another person's mind. Now, they may have good intuition and empathy, but in the long run relationships will deteriorate if mind reading is a major factor in communication. After a while, things left unsaid or unexplored create a fog around relationships in which the protagonists drift apart. Furthermore, mind reading is often an escapist tactic to avoid an unpleasant airing of differences. The above statement could relate to someone with a Sun-Saturn-Neptune combination in the 11th. It could be tackled by:

Astrologer: Exactly which people in the group think you take up too much time? [Asking the client to specify]

Client: I think they all do.

Astrologer: Precisely how do you know what all these people are feeling inside?

The client cannot know. Participants in group work are often convinced they take up too much time, but this tells more about the participant's own sense of identity than about the group, and that is the core issue behind the mind reading. Nevertheless, it can be extremely difficult to get people with a strong Neptune to accept that they cannot read another person's mind. They pride themselves on being able to. A typical statement could be, "I couldn't tell her, because she'd feel bad." Here, we have cowardice masquerading as consideration, so:

Astrologer: So, by not telling her, you are making her feel good? [Turning the sentence around to its converse statement]

The point is, in the long term this person's partner is never really going to know what's going on. Wrapped up in Neptunian cotton wool, there will be nothing to hold on to.

Mind reading is often pure projection. Take the statement, "If I tell my husband I want to go on the trip [to a Paris trade fair] with my boss, he's bound to get upset." If you are looking at a woman with a Mars-Neptune opposition, with which there is always the risk of succumbing to seduction attempts, it is likely that she is projecting an awareness of her own weaknesses on her husband, who may have no such inclinations or suspicions. You could turn this around with:

Astrologer: If he asked if you'd mind him taking his secretary on a trip, you'd get upset?

Client: I'd be beside myself with worry!

Which would naturally lead you to the dangerous fantasy world Mars-Neptune can create. Unless she checks it out, she'll never know. And if she doesn't, she will become a victim of her self-destructive imagination.

Counseling in Action

Let's look at how you can use the process of questioning to reveal the deepest issues in the following excerpt from a consultation. Bear in mind that before this process starts, it is essential to have won the confidence of the client and to have demonstrated the effectiveness of astrology. Using questioning techniques works best when the client thinks you already know the answer!

During the course of this example consultation—which took place in a group training environment—the person in question was prone to making sweeping statements with absolute conviction, as befits a Leo with Sun and Mars in conjunction:

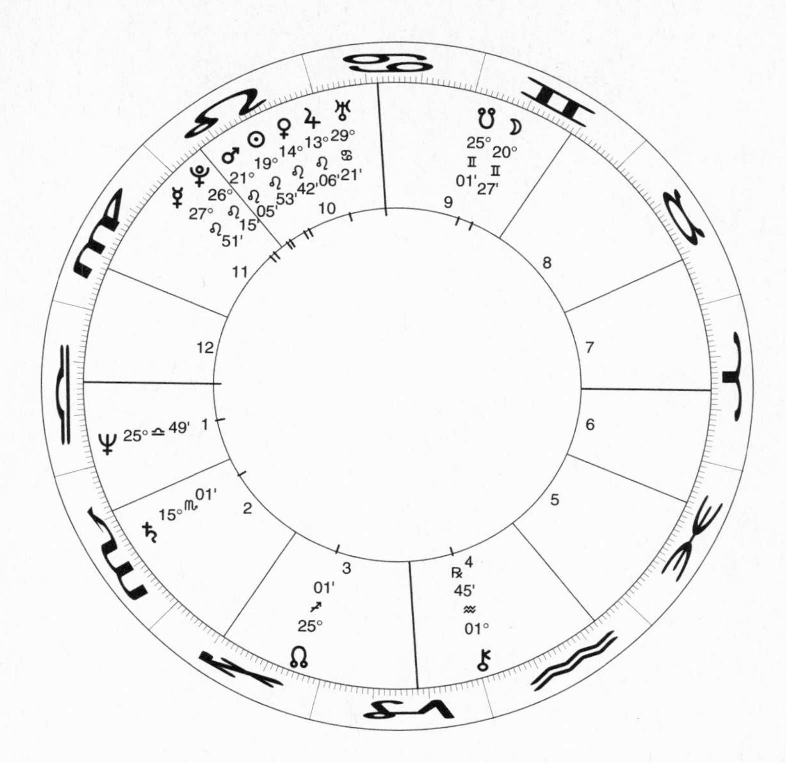

Fig. 5. Anthony. Views about history reveal deeper personal issues.

Anthony: Throughout the course of history it has been established that, in the long term, the majority is always wrong.

A wonderful assertion! It was amusing to see how the "majority" almost seemed to be reflected by the vast stellium of six planets in Leo. And the dark side of Saturn in Scorpio was determined to make them all wrong. What drama was this? Linguistically, this is a grossly ill-formed sentence. We don't know what "the course of history" refers to, who established the supposed fact, how long "the long term" is, who "the majority" are, or whether "always" really is true.

This kind of assertion, with words like "majority" and "history" are typical of 11th-house placements. Let's start on the feast:

Astrologer: Established by whom, exactly? [Getting him to specify "it has been established"]

Anthony: I don't know who specifically . . . it's my experience, anyway.

Astrologer: So it's your opinion then.

Anthony: Yes it is.

Now, although there are other issues here, the majority being wrong seems to reflect an experience Anthony has had, so let's home in on that:

Astrologer: Whom do you mean when you say majority?

Anthony: You know, consensus, the group, people who band together. [Nominalizations]

Astrologer: In your life, which group specifically has been wrong, and about what?

Now Anthony had to give an example from his personal life, and it was interesting. He was a member of a housing co-operative that decided to replace the windows in his building with new windows, while Anthony had argued for a restoration of the old windows, as the old wood was of so much better quality than new wood. The economy of this restoration process is obviously reflected by Saturn in Scorpio in the 2nd.

Astrologer: So how do you know they were wrong?

Anthony: Well, two years later, they admitted that the quality of the new windows did not compare to that of the old and that they only had a life of about 20 years, whereas the old ones had lasted 150 years.

Astrologer: So the housing committee agreed in the end that your suggestion was better.

Anthony: [With profound satisfaction] Yes!

Astrologer: So *in the long term* they agreed with you?

Anthony: They did.

Astrologer: So in the long term, were they right or wrong?

Anthony: Ah!

This is not trivial, because Anthony's sweeping statement reflects an attitude of confrontation in society that could cause severe trouble and alienation. Behind such a statement is a body of experience that is founded on a severe early trauma reflected by the Sun's configuration with Mars and Pluto and the square to Saturn. Whenever the Sun is involved, there is likely to be a connection to the father:

Astrologer: So, did your father tell you of any experiences where the majority was in the wrong?

And the story comes out: his father was in bomber command during World War II, and at some point balked at taking part in bombing raids on the German civilian population. As a result, he was sent off to a distant colonial outpost, and there were no further promotional developments for him in the Royal Air Force. It is fascinating to see how the enormous destruction of places like Dresden seems reflected in Anthony's chart—both the economic destruction (2nd house) and the elimination of the population (11th house). The important thing for Anthony, however, is the social repercussions that he absorbed from his father's stories and attitude, and how they have subsequently affected his own integration into society and ability to perform in his career.

With practice, recognizing what lies under the surface of language structures becomes instinctive. It's simply a question of asking yourself if you have fully comprehended everything that the client has said. If you have not, then you should investigate. You may feel that this breaks up the flow of conversation—and it does—but there is no faster way of getting down to the hidden issues that are preventing the client from living a fuller life. There is no doubt that this form of questioning is intrusive, and normally there is no need for it to last more than a few minutes. After this time, you will have reached a core issue, and there is no point in continuing to tramp around on it. You will have to use other, gentler techniques to resolve the deeper issues. This is crucial. If you persist in these questioning techniques, getting deeper and deeper into what could be a traumatic state, your client will end up feeling bad. In the next chapter I'll show you how to put the client into a positive, resourceful state before you embark on this process, and, in later chapters, we'll look at how to help the client resolve core issues.

3 Evoking Inner Resources—Empowering the Client

Whatever opinion you may have of former president Bill Clinton, his remark that "There is nothing wrong with America that what's right with America won't cure" encapsulates the essence of the importance of work regarding accessing resources in the individual. There is nothing wrong with a person that what's right with that person won't cure. In other words, every individual has the resources within to handle almost anything that turns up in life. However, it is human nature to focus too often on problems and difficulties rather than on strengths and successes. Moreover, astrologers are inclined to focus on astrological configurations reflecting difficulties, rather than those reflecting happiness. There is a very good reason for this, and that is that the client tends to give a far clearer feedback about difficulties than about strengths. People come to counselors to talk about problems, not to be reminded of what is good about themselves. However, what both astrologer and client can fail to realize is that without the necessary resources, they will make ineffective progress solving personal problems.

The Importance of the Positive

It should be clear that pain and difficulty are reflected by the traditional negative aspects: oppositions, squares (and semi/sesquisquares), and conjunctions to the traditionally difficult planets, primarily Mars, Saturn, Uranus, Neptune, and Pluto. There is a school of thinking that celebrates difficult aspects, because they are where growth and dynamism lie, and, of course, this is true in the long term. It is possible to transform difficult aspects into resources. However, these difficult astrological configurations will unfailingly bring problems early in life and, indeed, for many people, all the way through life. Aspects indicating resources—"soft" aspects—are trines and sextiles, as well as conjunctions to beneficial planets, particularly Venus and Jupiter. Strongly placed planets (for example, exalted, or in their own sign) also indicate resources. Good influences like these are not just positive, they are unique and magic blessings for the client, which, when experienced and recognized by the client, evoke a sense of gratefulness and privilege.

Twentieth-century attitudes to soft aspects have been rather dismissive. Astrologers do not doubt that they bring benefits, but some schools of thought marginalize positive aspects as being related to comfort, ease, and weakness. Indeed, in branches of midpoint astrology, the positive aspects of the sextile and the trine do not register, as only divisions of the circle by factors of two (rather than three) are seen to produce events, or results. Of course, each number has its quality, as harmonic astrology shows, and the quality of three is harmony, joy, and stability. It is unwise to exclude these states from the astrological consultation.

But how do you describe a good influence to a client? What will often happen is that just a few minutes can be spent elucidating a good aspect, while it is easy to spend half an hour talking about a difficult one. So, in the case of Sun trine Jupiter in fire signs, for example, it might go something like this:

Astrologer: You have a wonderful influence in your horoscope—a beneficial contact between your Sun and Jupiter—which shows you as a dynamic and positive person, with a deep understanding of, and zest for, life!

Client: Hmmm . . . well, I'm certainly interested in travel and philosophy. People do say I have a positive attitude.

Astrologer: Right. It means that whatever difficulties arise, you are buoyant, or at least you bounce back very quickly.

Client: [Who has, say, Pluto transiting Moon in Scorpio] Well, I don't feel very buoyant right now . . . I've been down for months.

Astrologer: [Shifting to the Pluto transit] Well, that's not so surprising right now, because . . .

Time and again in the consultation, astrologers are distracted from soft aspects, or give up on them, because of other issues that seem more pressing. To work with resources the astrologer must resolutely press on elucidating the positive aspects; there will be time to work on the difficult aspects later.

Delineating soft aspects is one of the most difficult arts in the astrological consultation, but the rewards are stunning. To do it successfully requires intelligence, inventiveness, and sheer obstinacy. But it also requires a clear understanding of sensory states, and this must be learned, and the skills to evoke these sensory states must be acquired by practice and training. The secret of resources as shown by the horoscope is not to describe, but to *evoke*. To do this means inducing an energy state that exactly accords with the pure experience of the harmonious aspect. As far as I know, this cannot be done without the astrologer at least partially entering the same state. And why not do this? It's a very nice experience, indeed. All soft aspects have sensory states associated with them—in fact, they *are* sensory states. A Sun trine Neptune successfully evoked can actually trigger the same original spiritual experience that it designates. Similarly,

Venus in Taurus sextile the Moon, successfully evoked, will bring an inner-body sensation of intense pleasure. Evoke Mars in Capricorn sextile Saturn in Scorpio and the client will feel a surge of power and executive ability and feel like Master of the Universe.

In evoking these states, it is crucial to make no value judgments whatsoever. For example, Venus in Scorpio in the 5th, trine Mars in Pisces will likely refer to taboo sexual experiences that could best be described as noncatholic, but the point is that these experiences reflect a sensory state and that this state represents an extremely powerful resource. How the person feels is more relevant than what the person actually does. When dealing with resources that may have a secret character—which erotic experiences have—it is not even necessary to know precisely what events and actions correspond to the aspects. All you have to do is insure that the sensory state is being experienced strongly.

The point of evoking a powerful and pleasurable sensory state is that it is impossible for a person to experience such a state and still feel bad. During the time the person experiences the state, he or she is going to feel really, really good. For some, this is therapy enough. But the important thing is that the client can now approach a problem from a position of strength. Acute problems simply cannot be solved without resources. Attitude is everything, and with a positive and enthusiastic attitude, nearly intractable problems can seem to dissolve, whereas even the smallest difficulties threaten to engulf a person with a negative attitude and low valuation of resources.

The Positive Aspects

To commence the practice of mobilizing resources, choose the strongest beneficial influence you see in the chart. The strongest beneficial states are shown by trines, which in essence are passive states. The influence of a trine can best be described as a deep and clear well, which will always yield the pure water of beneficence whenever the user chooses to draw from it. It's there as a continuous source, and it's possible to access the associated sensory state at any time. It never leaves the client. Sextiles also evoke positive states, but there is a precise activity or effort associated with them, probably related to the numerical fact that both polarity (number 2) and ease (number 3) are contained in the number 6. When working with sextiles, the activity or effort becomes extraordinarily clear, for the client will precisely describe the influence and tell of the effort involved in attaining the pleasure.

On very rare occasions, you may see no remarkable trines or sextiles in the client's chart, in which case you can use a strong planet as the vehicle for evoking a positive sensory state. I find that the traditional concepts of exaltation and fall work really well, with Sun in Aries, Moon in Taurus, Jupiter in Cancer, and Mars in Capricorn particularly amenable. (Venus in Pisces definitely relates to a high spiritual state, but see the following section on water signs.)

Planets in fall are ill-suited for initial work with positive aspects, because when clients associate with them, they do not immediately think much of the experience. Jupiter in Capricorn is something of a moralist and slogger, for example, with painful memories of unjust authoritarianism, while Sun in Libra experiences identity subordinated to others, and Moon in Scorpio has many traumatic memories. Because you are evoking sensory experience, it can be dangerous for the client and alarming for the astrologer working with these energies—even when they are part of a trine or sextile—but it's not impossible. If you are tempted to try, in the case of a Scorpio Moon for example, you'll find the client accessing something deeply painful. The trine shows some ease in dealing with it, but the memory hurts. You'll be led into uncharted waters, so be prepared. Planets exalted or in their own signs will always bring positive results.

The Relative Strength of Resources

When working with resources you should look for planetary constellations involving personal planets—preferably the Sun, Moon, Mercury, or Venus. They can constellate with themselves (for example Moon sextile Venus—nice!) or with the planets beyond Earth.

It's great when Jupiter is involved, because it will always bring pleasant associations.

As you experiment with evoking resources you may find it difficult at first, so it's best to choose something obvious and easy. Aspects from inner planets to Uranus and Neptune are somewhat more rewarding to work with than good aspects to Saturn and Pluto. This is because any aspect to Pluto will be associated with a trauma—the resource is the sense of survival of that trauma and positive attitude to future threatening situations. Any aspect to Saturn will be associated with a great difficulty, but a difficulty often successfully overcome. The resource associated with Saturn is the sense of accomplishment and determination to reframe problems as challenges. Neptune, too, has its pitfalls: even positive aspects to Neptune are associated with longing and sadness, but the resource associated with this is a wonderful, transcendental acceptance. I have not noticed clients having any unhappy associations with soft aspects to Uranus.

In summary, don't make your life difficult—choose an aspect with the best chances of bringing the client into a strongly charged, positive emotional state. Jupiter placed strongly trining, sextiling, or conjoining (in that order) is best. Jupiter brings associations of space, panoramas, and vistas, particularly in fire signs. There is a sense of joy, exaltation, fulfillment, and, above all, a spiritual sense of deep understanding, of knowing. Venus is next best, basically because of the strong and pleasurable body sensations, particularly in earth signs. With Venus, there is a sense of happiness, harmony, balance, and pleasure—a true sense of being blessed. So, if you can find a Venus-Jupiter combination, you simply cannot go wrong.

One of my students, Janne S., has Venus in Taurus (strong in its own sign) in the 7th, sextile Jupiter in Cancer (exalted) in the 9th (see figure 6, p. 59). What more could you ask for? It's true that she also has Sun-Mercury in the 8th, square Uranus-Pluto in the 11th. However, what's right with her will give her the ballast to sort out what's wrong with her. It's interesting that she worked at one point for a law firm (Jupiter in the 9th), and her greatest pleasure was

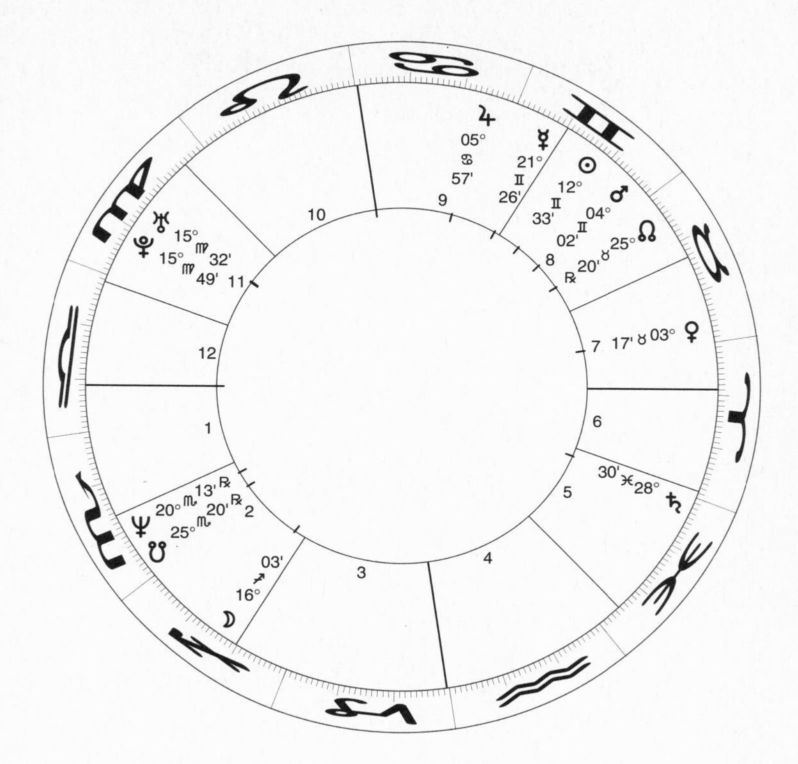

Fig. 6. Janne S. A strong Venus and an exalted Jupiter bring the blessings of riches, intelligence, and influence.

---

in making wonderful dinners for them (Taurus). She received great praise for her culinary skills, which, of course, were in the cordon bleu domain. Her parents were very, very rich, and she remembers the feeling of driving around in the white leather interior of their Mercedes. She claimed at first not to have enjoyed this, because she had such negative associations with her father's wealth. In other words, during the exercise of evoking a positive sensory state, she got side-tracked by the energy of the square from the 8th to 11th houses, and Neptune in the 2nd. Whatever her aversion to the brazen social statement of a luxury Mercedes, the sensory pleasure of the leather interior and the sheer smoothness of the drive hooked in to her Venus-Jupiter sextile and was stored as a positive sensory experience, just as her law colleagues' affirmations gave her an exquisite sense of being appreciated. This feeling, coded into every cell of her body, is a match for whatever negativity the difficult aspects in her chart have to offer. In essence, neither the praise of her superiors nor the purr of the Mercedes is what is important here; what really matters is the sensory buttons that get pushed by the experience.

The Elements and the Senses

We can go no further without an understanding of the nature of sensory experience. All pure experience is filtered through one of the five senses: vision, hearing, touch, taste, and smell. As soon as conceptual filters are applied, the individual takes a step away from being in touch with reality. When the rational mind is active, then experience is filtered at best through a set of opinions and convictions and at worst through a warp composed of a tangle of memories and problematic attitudes. Accessing resources means completely bypassing the rational mind and then accessing the sensory state.

Sensory impressions are stored according to the nature of the five senses. Pictures, sounds, sensations, olfactory, and gustatory experiences constitute hubs of regenerative power. Plugging into these power sources requires identifying the astrological resource, explaining it to the client, inducing the sensory state, and generating the power. The crucial factor in inducing these states is to realize and act upon the fact that each sensory state is associated with one of the four elements: earth, air, fire, and water. I have not come across any astrological literature that has specifically made this connection, although traditionally these elements were called melancholic, sanguine, phlegmatic, and choleric respectively—words somewhat associated with these senses.[7]

Practitioners of NLP certainly realize the significance of four "representation systems," and in this field, it is often maintained that people are inclined to use one representation system more than

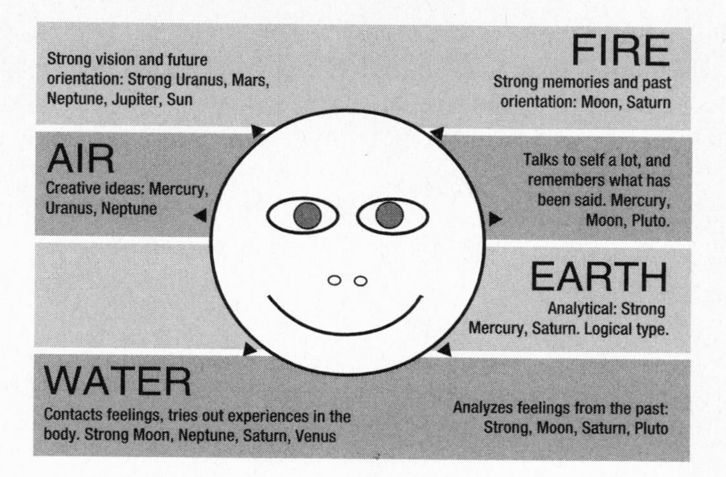

Fig. 7. An astrological model of sense filters. Sensory preferences are reflected by the whole body, but specifically by eye movements. Coupled with an understanding of how planets modify sensory preferences, it is possible to read clients' signals and proactively lead them.

another. However, by using the horoscope, you can be far more specific, and it is possible to see which sensory system is preferred for what specific activity by studying the type of aspects and planets involved, and the elements they are in. In other words, people will use one sensory system when involved in activities symbolized by, say, fire signs, and another when involved in activities symbolized by earth.

In NLP, these four systems are termed visual, auditory, kinesthetic, and audio-digital (the last being a kind of "Mind" sense, which I would associate with aspects of Virgo, Mercury, and perhaps Mercury-Saturn or Mercury-Pluto combinations, that is, accessing the past, having an inner dialogue, or being very rational). Of course many systems use four pillars as representing four very significant states, and I am not suggesting that the four astrological elements exactly translate to these states, just more or less so. Transferring meaning from one psychological or philosophical system to another is hazardous, and the diagram above (figure 7) should be viewed more as a guide to inspiration rather than chiseled in stone.

This diagram of the head and eyes is a way of illustrating what sensory state the client is accessing at any particular time, and students of NLP will notice a number of similarities with eye-movement maps used with NLP techniques. Practitioners use the direction of eye movements to confirm which representation system the client accesses most. Eyes upward indicates visual, eyes horizontally to one side or the other indicates auditory, eyes down to the left, auditory-digital, eyes down to the right kinesthetic. In practice, people who look down to the right access feelings, and down to the left thoughts. People who look up to the left access pictures that they already have in memory, and up to the right they create pictures (visualize). Those who look horizontally to the left access sounds they have heard, and those who look horizontally to the right imagine sounds they have not yet heard. In these cases, the left-hand side seems more about the past and the right-hand side more about the future. Just to make it more complicated, there are exceptions to this scheme of things, and some people have eye movements reversed. Others—for example those with strong Scorpio or Pluto influences—instinctively control eye movements to reveal little. In practice it can be rather difficult to use eye movements consistently in therapy.

This mapping of eye movements and relating them to sensory representation systems was an extraordinary discovery by the early pioneers of NLP, and in the consultation process—given that facial expression and eye movement is a key factor in interaction between you and your client—you can use it to advantage. The diagram shown here is an adaptation of the original idea using astrological elements and their sensory counterparts, rather than the sensory categories of NLP. The following section explains how it works.

The Fire Element

People with resources related to planetary configurations in fire tend to look upward when brought in touch with this energy, and the shading in the top of the diagram shows that fire energy is at the top, just as the energy of fire could be said to rise upward. Because fire has a visionary quality, it gives the capacity to look forward in time and visualize a scenario in the future, as well as clearly see a situation from the past. These people are constantly creating pictures that are very real for them, like movies, three-dimensional and in full color. This talent of visualization can be used very successfully to induce a positively charged sensory state. Often planets in fire are experienced as pure energy, sometimes as light both outside the body and permeating the body. It's an energizing experience, bringing warmth, power, and a very strong sense of identity.

People manifest fire resources by holding their body upright, with their head inclined at an upward angle, often looking toward a distant horizon and often using their upper body and arms to express themselves. Significantly, they use a vocabulary connected with "seeing," talking about pictures, visions, sight, glimpses, and view. When they understand something, they say, "I see." They talk animatedly, and with obvious enthusiasm. Therefore, it is important, when trying to get a client to access resources connected with fire, to match the client's vocabulary, tempo, and body language. When you are working with resources, it is often the case that the person is not at all in touch with them and is focused on a negative experience of some sort, so getting them to feel positive requires considerable skill. As you describe a trine aspect in fire, you have to imagine you have it yourself, and adopt its vocabulary, tempo, and body language. If you do, the whole process happens much quicker; if you don't, it may take longer or never happen. It would simply not work to slump in your chair and start asking questions in a laid-back tone. You have to imagine being possessed by the energy, and it's OK to exaggerate a little.

The activities that the fire element likes to engage in are exploratory and adventurous—activities of excitement and daring. Often they will involve initiation and new understanding. Because the first step in evoking a resource is to identify the particular activity associated with the resource, this is where your investigation would start. You have to find a specific event and expand on

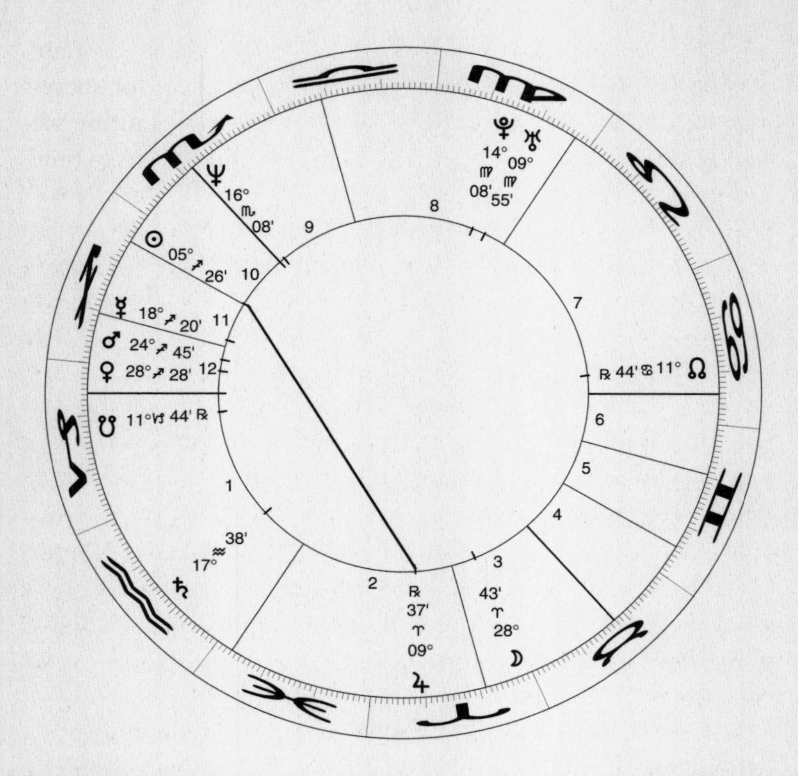

Fig. 8. Gitte D. A strong Sun and Jupiter in fire, giving great powers of visualization.

the sensory experience. Mars in fire will be concerned with risk-taking and competitiveness, and the resource lies in the buzz of winning. Venus would be more associated with romantic excitement and the winning of hearts. The Moon might relate to an immense enthusiasm and warmth, the Sun to a supreme satisfaction with one's self. Mercury would be associated with mental competition, travel excitement, storytelling, fun, and games. Jupiter would be connected with impressive views, space, expanse, and deep understanding. These would be the planets most amenable to resource work in fire.

Fire signs do love winning, and they also like to be right. Having positive aspects in fire signs can give a talent for success because of the instinctive capacity to visualize positive future scenarios. While many business courses strive to train recruits to think positively, this come naturally to the fire sign person. One client of mine is one of the world's best female backgammon players and makes her living solely through competition playing. She had a technique for winning that involved visualizing herself standing on the podium after the contest receiving the cup and considerable cash prize. She could see the participants looking at her, hear them clapping, and feel the warmth of success lighting up her body. Her Sagittarian Sun on the 11th-house cusp, trine Jupiter in the 2nd house (note that Jupiter rules Sagittarius and the Sun is exalted in Aries, so there is mutual reception) obviously reflected her extensive travels and gambling career, but her ability to win was founded on her visualization strategy. When this state was evoked in her, she felt tremendous empowerment, and this empowerment enabled her to develop a positive attitude to the more difficult areas of her horoscope (see figure 8).

When the client connects to the feeling of the fire resource, a noticeable color will come to their cheeks and neck, and there will be a sense of awe, power, and excitement, as a strong self-awareness dawns. It will often be the case that the client has not been conscious of this exhilarating sense of identity before, hence the awe. There is a feeling of greatness that suggests that there is something above and beyond identity and, via this feeling, a connection to a higher, spiritual state.

The Air Element

People with resources connected to planetary configurations in air signs are very rational and objective, often talking about perspective, distancing, inspiration, and stimulation. There is a calm and clarity about them, and a strong sense of mental overview. Their vocabulary is dominated by auditory concepts, using, for example, the word "hear" to signify understanding and phrases like "it sounds like" when they work with ideas. They experience things as being in "harmony" or "discordant," or they "tune in." As the main reference system for resources connected with air is auditory, they access air resources through hearing words and sounds internally.

Let's say someone has Venus in Gemini in the 9th, trine Mars in Libra in the 1st house. Their sensory associations connected with the trine would have to do with someone's voice, or words spoken to them, as well as the whole environment they are in. An obvious event connected with this aspect could, for example, be a romance with a lecturer. Associations with his or her voice, any letters, any trips together, would be powerful experiences in which the resource and the enormous pleasure associated with it come to the fore. Although Venus-Mars combinations are often erotic in nature, with this person, intellectual seduction would be important. Alternatively, the whole social and intellectual environment of a college would give such a client an extraordinary sense of happiness and blessing.

It can be difficult for the astrologer who does not personally have much representation in a particular element to help a client access that element. This is because there are no instinctive channels at the astrologer's disposal to tune in with. However, with the use of appropriate vocabulary relating to the planets and signs involved, and a determination to find just how a particular aspect manifests, it is normally possible to guide the client in the right direction. I recall a rather difficult consultation with a Libran, whose 12th-house Sun was trine Uranus in Gemini in the 8th house (see figure 9, page 67).

Strong air-sign types can seem abstract and remote to me at the best of times, and I had difficulty establishing a rapport with this man, who worked as a pilot for an air freight company. His plane carried crates, not people, and I am sure this suited the private inner nature of the 8th and 12th houses. With a crew of just a few people—as a Libra he did have his team—he spent hours at a time and decades of his life on long transcontinental flights. It's easy to see this reflected in the Sun-Uranus trine, but where's the resource?

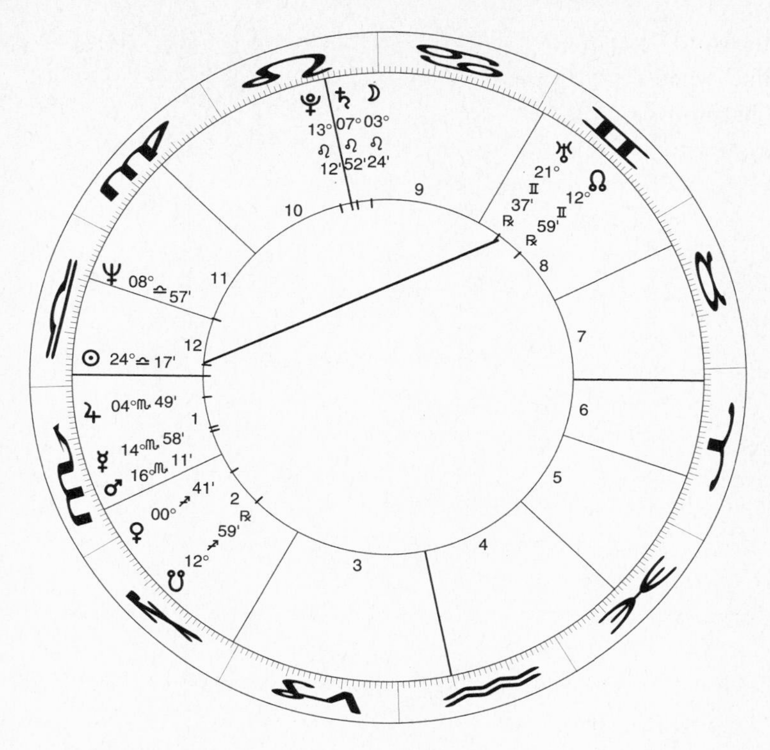

Fig. 9. Gunnar B. A trine from within the secretive 12th and 8th houses brings deep spiritual joy.

Astrologer: Is it boring spending all those hours at the controls?

Client: No. [He was a very monosyllabic man.]

Astrologer: How is it interesting?

Client: Away from it all. Above the clouds.

Now, linguistically, we would want to know just what it was he felt he had gotten away from, because therein lies an opportunity for a therapeutic intervention, but to do this would lead away from the resource so:

Astrologer: How does it feel then to be jetting through the stratosphere?

At this point he shared the profound feelings of joy he experienced in flight. For him, it was a spiritual experience. He felt in touch with the gods. Flying was his life, and this was why he had come to see me—he was due for compulsory retirement and had no idea how to cope with the imminent removal of his greatest source of pleasure. I was able to get the message through that you don't need a plane to fly on the wings of Sun-Uranus. His need for spiritual stimulation could be satisfied by a variety of other pursuits.

Emphasis in air is very refined, and very mental. It is often not necessary for this type to achieve anything concrete to feel pleasure. Sometimes the act of mental gymnastics, the play of ideas, and experimentation is enough. It is not easy to capitalize on abstract mental resources in the consultations, so it is important to identify the outer manifestations of these inner states. They are chiefly to be found in an active social life and in communication. By focusing on the experience of friendship and the sound of voices the person loves, an empowering sensory state can be accessed. When planets in Libra are involved, there is often a strong appreciation of aesthetics. Art, writing, and beauty in all its manifestations can be the triggers that create an inner sense of harmony, which can be utilized as a resource.

As with fire resources, air influences manifest in an upright and animated body posture, with a considerable amount of gesticulation. Eye movements would also be in an upward or horizontal direction, and there would be a good degree of eye contact, as the sharing of ideas and the need to interact socially is strong. Memories of being stimulated through learning and from being together with specific people would be associated with air resources. Working as an astrologer to access these resources, you would need to communicate in an animated way and concentrate on sounds when evoking the resources: the bubbling of conversation, words of love, the exact sound of a person's voice. When people connect to the energy, there is often a marked change in the tone of their voice, and a rather breathless sense of excitement.

The Earth Element

There is a considerable difference when dealing with resources connected with the "feminine" signs of the earth and water elements. These resources have a heavier feel, and the client will be far more in touch with his or her body. When working with resources in this way, it's easy to see how accurate traditional astrological thinking has been, for the resources associated with earth have little to do with the vision of fire, or the mental stimulation of air—they are so obviously concerned with practical achievement and physical pleasure. If Taurus is strong, then so is the pleasure principle; if Virgo or Capricorn is strong, then pleasure of achievement is also significant.

As a means of accessing resources, the earth element is by far the easiest to work with, because it is associated with body feeling. As a matter of fact, *all* resources come down to body feeling in the end, so even when working with fire or air, it's important to get the client to locate the exact body sensation associated with the resource. Fire people will feel a strong warmth and often talk of a sensation of aura; air people might sense tingling and surface sensations. The body feeling of earth is tremendously strong and all-pervading, and very specific in terms of the planet and sign involved. It's amazing to hear the client talk of pleasurable sensations in the feet with Venus in Pisces, in the throat with Venus in Taurus, or in the heart with Venus in Leo. It is a revelation to learn how accurately the twelve signs correspond to body areas in many instances. Naturally when working with a client, it is not a good idea to prompt or lead with questions like, "Do you feel a sensation in your heart?" It is far more satisfying and educational to let clients describe in their own words how and where they feel a sensation. However, after a little practice, you will simply *know* where they feel the sensation.

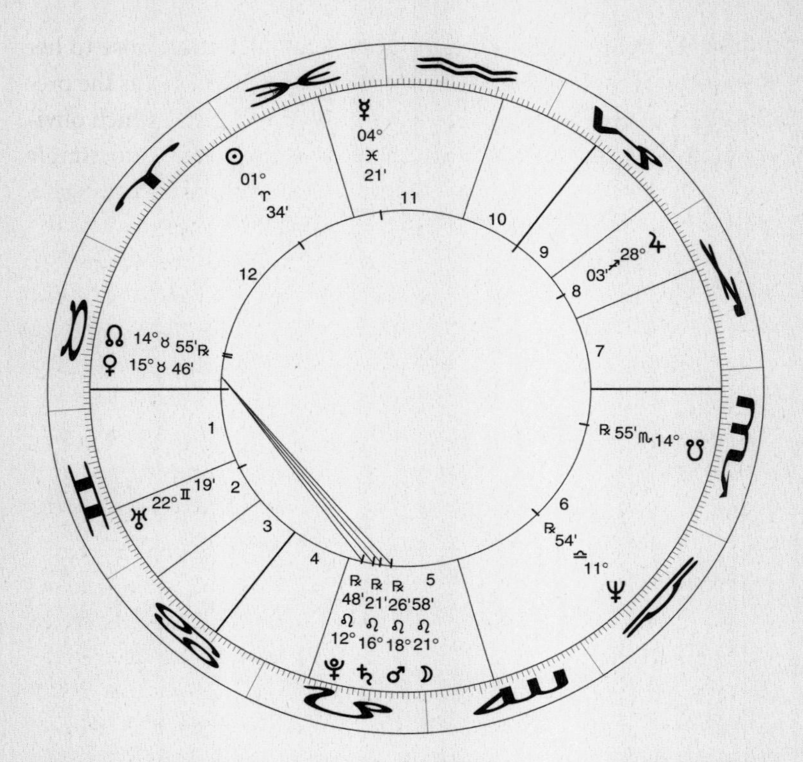

Fig. 10. Birgit K. Despite the difficult aspects to Venus, its position as horoscope ruler, in its own sign, gives great sensory resources.

One client had a Taurus Ascendant, with Venus in Taurus in the 12th (see figure 10). There were no soft aspects in her chart at all, and it was dominated by a Moon-Mars-Saturn-Pluto stellium in the 5th house—not exactly fun. As this woman was a yoga teacher, it was obvious she was strongly in touch with the Taurus energy and her body, so I chose to evoke resources via the Venus in Taurus in the 12th, which represented an inner, almost spiritual sense of body feeling.

By getting her to access the body sensations in different parts of her body, just as she might do in a yoga lesson, she was able to feel an extraordinary empowerment. As her body sensations rose to her hip and groin, she then declared she was blocked, which was the precise effect of the stellium in Leo square Venus in Taurus, which obviously also related to sexual blockages. The trick here is to simply ignore the blockage and go on, though one is tempted to investigate. The golden rule of resource work is to resolutely ignore the client's desire to relate problems and blockages, and to work on them later. If you get distracted, you will be led astray, and the client will lose contact with the resources she was about to access. Clients will often say, "Yes, that's great, but . . ." You may take note of the "but" for later attention, but otherwise you should bypass it completely.

When clients access their earth resources, they tend to experience a particular body sensation, especially if Taurus is involved. There is none of the gesticulation and excitement of fire and air. Few words are spoken. The body is neither taut nor erect, but very relaxed, exuding pleasure. The job of the astrologer is not to disturb too much, but to ask questions expanding on the physical sensations experienced by the client. It should be clear that the point of asking the questions is not because you need to know the answers, it is exclusively to induce a stronger and stronger sensory experience in the client. The stronger the experience is, the greater the empowerment.

The Water Element

If you are going to work with water resources, you should have your box of tissues ready, and not just for your client. Water resources are powerfully connected with the emotions, and there is almost always a tinge of sadness connected with them. Indeed, it can appear that the client gets very unhappy when accessing this energy, although this is only a matter of interpretation. They may think they are sad, but the energy is very cathartic. It is perfectly possible to be sad and deeply moved at the same time. As I mentioned before, working with resources works best when the astrologer is completely tuned in and feels corresponding sensations, so you have to be prepared to enter a painful and poignant world when dealing with water.

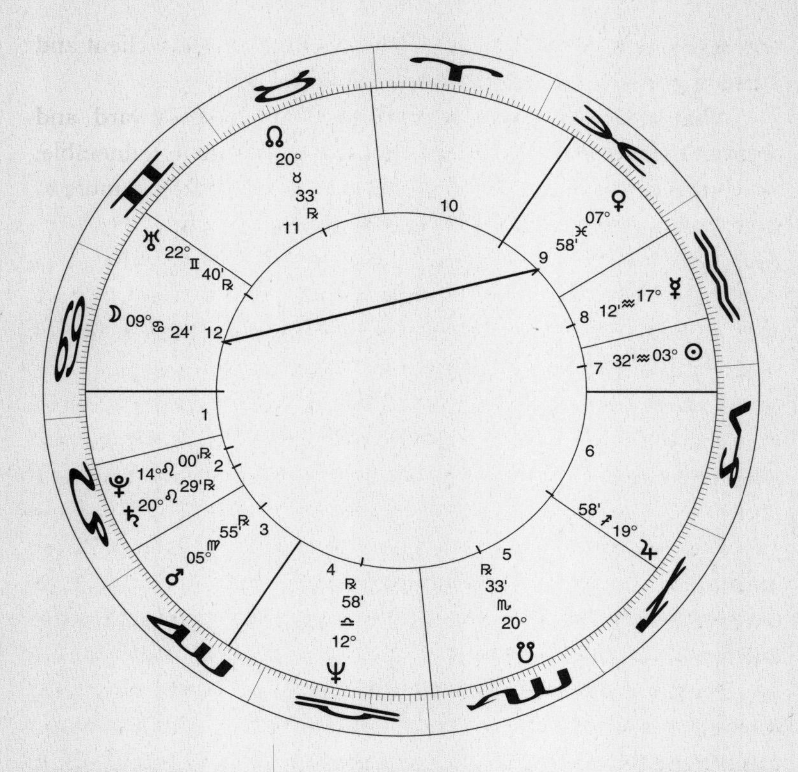

Fig. 11. Lene V. A strong Moon and an exalted Venus make a magnificent water trine, full of joy and sadness.

In the Element correspondence diagram in figure 7 on page 61, water has been placed lower than earth, which is satisfying symbolically, for water will always seek the lowest reaches. Nevertheless, the diagram should be understood more as fire and air directing attention upward and outward, and earth and water downward and inward, and in this scheme of things, water shows what is deepest within. Water is the most challenging element to work with because it creates a very intimate atmosphere. Clients are often not willing to expose themselves emotionally, and astrologers may also have a similar reluctance to show their emotions. But the nature of emotion is that it is pervading, and in the consultation both client and astrologer will experience the same emotional field.

What normally happens is that clients look upward and abstract themselves from an emotion that makes them vulnerable, which makes it impossible for them to feel it fully. Take a minute to experiment: get in contact with an emotion and look upward and to your right, and try to still feel the emotion. Now look down and to your right. You should notice quite a difference. So, when getting your client to access water resources, speak softly, gesture down to their right, and create an environment of expectant silence.

If Neptune or Pisces is involved in a water trine, then the sadness and longing is considerably intensified. In fact, Neptune will give something of the same poignancy in any of the four elements. One of the most beautiful examples of a water trine I can recall was a female client with Moon in Cancer in the 9th, trine Venus in Pisces in the 12th (see figure 11 on page 72). How would it be possible to have a stronger and more beautiful water trine, with Venus exalted and the Moon in its own sign?

Her story was as follows: This woman worked with her husband, teaching (appropriately) healing and meditation. Curiously, she came to me because she was having a problem selling a house, which had previously belonged to her mother who was now dead. Naturally, dealing with the house brought back a lot of family memories. The mother had been in a hospital in another part of the country when the client received a message that her mother was dying. She rushed to the airport and arrived at the hospital just in time. Fortunately, her mother was still lucid, and when she spoke to her daughter as she leaned forward—not having much control of her body functions—she ended up accidentally spitting in her daughter's face. "Mum, you're spitting at me!" exclaimed the daughter jokingly, at which both mother and daughter laughed hysterically for many minutes. The death took place soon afterward, while the daughter stroked her mother's feet. Trines in water signs—get it?

When clients access water resources, therefore, a lot of patience and sensitivity is required. They may be reluctant to go so deep, so to make it work, the atmosphere must feel very secure. When you see their attention drawn upward and hear too many words, you are on the wrong track. People may well access the event via seeing and hearing, but each image and sound will strike a chord deep within. When you sense the client registers this chord, then you should speak only to further enhance the client's sensation. It could go something like this:

Astrologer: I noticed you were feeling something rather strong then?

Client: Yes. [Silence]

Astrologer: Where specifically in your body is that feeling?

Client: In my throat. [Silence]

Astrologer: As a kind of pressing feeling, or warm feeling . . . or what?

Client: It's a kind of open feeling. [Silence]

Astrologer: Which makes you feel how . . . ?

It might seem rather silly to ask how a feeling specifically makes a person feel, but it actually makes the person get the feeling behind the feeling. All feelings are seated somewhere in the body, and it is essential in all resource work with earth and water to come down to a basic body sensation and expand on it. The questions might sound banal, but this is an effective technique for inducing a stronger and stronger feeling. The client's words and answers are not so important at this stage, and the last thing you want is some rationalization—which would signify that the client was getting detached from the feeling. It serves no purpose to rationally understand what is going on. When working with water in particular, you are in fact inducing a trance in which the client goes into a meditative state, and, when successfully done, it is a moving experience. Figure 12 (see page 75) shows what kinds of questions are appropriate for which elements. Note too, that certain planets enhance certain elements and senses. As you ask questions, your client accesses the sensory experience

VISION - FIRE	HEARING - AIR	BODY - EARTH	TASTE/SMELL - WATER
1. Where? Distance? Angle? 2. How? Color or black and white? Movie or stationary? Size? Framed or pan-oramic? 3-D or flat? 3. Dissociated? Is self seen in picture, or is everything seen through own eyes?	1. From where? In your head or outside? Where from outside? From one point, or from everywhere? Sound comes in through your ears, or where? 2. How? Volume? Tone? High- or low-pitched? Monotonous or musical? Tempo?	1. Where in the body? How much space does it take? What shape is it? 2. How? Hard or Soft? Warm or Cold? Heavy or Light? Dense or Empty?	Sweet or sour? Salty? Sickly? Poisonous? Delicate? Bouquet? Pungent? Sharp? Sweet? Smells of what?
PLANETS Uranus, Jupiter, Sun Mars	Mercury, Uranus	Saturn, Venus	Pluto, Moon, Venus, Neptune

Fig 12. Evoking sensory experience. Ask questions to guide the client into a sensory experience of an event. This promotes memory, inducing a light trance from which resources and creativity are accessed.

internally, and with this internal access comes an experience of energy. It is this energy that relates precisely to the aspect pattern involved and that constitutes the resource.

Change happens at a deeply unconscious level when the client reaches this state: connections are made, unforgettable understanding dawns without words. The client becomes extremely open to suggestion, which is why everything has to be phrased delicately. If you have been talking about a problem, for example, you would want to refer to it perhaps as "the problem you had" (it's in the past), rather than "your problem" (it's still there). It's crucial to use language that assumes that things are being solved, rather than language that holds the client in the past.

Smell and Taste

Apart from seeing, hearing, and inner touch, the two senses of taste and smell are both important, and immensely powerful. Both earth and water seem to be connected with these two sensory systems, and they are closely interlinked. People who have a representation system powerfully associated with taste may well use expressions like "it leaves a bad taste in my mouth" or "that's not to my taste" or "juicy" and "spicy" to describe sensations and experiences. Taurus and Venus are particularly strongly associated with taste. To access this resource, you can simply expand on great culinary occasions or particular events concerned with food. Going back in time, taste is centered on the sensation of suckling at the mother's breast, and as such, it has a powerful effect to access this sense. Tense aspects to Pluto, especially from the Moon, often show a negative experience connected with breast-feeding.

Smell is even more acute and has an incredibly powerful effect. Smell bypasses the rational mind to go straight to the experiential core. Pluto is strongly associated with both good and bad tastes and smells, so good Pluto-Venus aspects will almost always indicate powerful sensory experiences connected with these senses. Among other things, this will give talents connected with things like perfume and wine. Often, though, when Taurus is involved, the sense of smell and taste is so advanced that these people dislike strong, artificial smells (as in most deodorants), and they will extol the scents of nature and the natural smells of the body. They love pheromones, and taste and smell will constitute an important part of their erotic values and experiences.

Dealing with tastes, and particularly smells, is a bit of a risky process. Any bad associations with smells are likely to suck the client into a traumatic vortex. Most stressful Pluto aspects to Moon, Mars, and Venus conceal really strong associations with bad smells that would undermine any effort to access a resource; however, when you wish to access problems, they are a powerful way in. It is often the case that people who use rather overpowering perfumes and deodorants have hard aspects to Pluto and a corresponding fear of "bad" smells. Interestingly, anxiety about smells was very dominant in the 1980s and early 1990s, when Pluto transited Scorpio, and there was a marked increase in men's use of overpowering deodorants at that time.

Another factor that makes accessing smell rather delicate is that some of the most resourceful associations connected with taste and smell are connected with sexual experiences, which can be a touchy subject to expound upon in the consultation. Nevertheless, if you get the client to access resources connected with the sense of taste or smell, then it has an immediate energizing effect, stronger than accessing any of the other senses.

Some Notes on the 12th House

When getting a client to recall an incident associated with a resource, the crucial thing is that the event be powerful. Normally, the earlier you go back into childhood, the more powerful it will be. There will be a string of events throughout life connected with the resource, and there is no point in finding out about them all. Choosing one and concentrating on the emotions and sensations connected with it is sufficient. In the example of Lene V. (page 72), the Moon-Venus trine basically symbolized an ongoing love between daughter and mother—the mother's death happened to be the strongest association at the time of the visit. With trines involving the 12th house, the bond will almost always go beyond death. In other words, there is an ever-present spiritual bond, which is still the source of empowerment.

Another example of this 12th-house theme was a client who had a Moon-Neptune conjunction in Scorpio in the 12th, sextile Mars-Pluto-Uranus in the 9th and trine Saturn in Pisces in the 3rd. This kind of soft aspect would be extremely difficult to work with as a source of strength because it has all the ingredients of trouble. The Moon is in fall in Scorpio, and its conjunction with Neptune and placement in the 12th shows a deep sadness. Good aspects to Mars, Uranus, Pluto, and Saturn do not show pleasure; they show battle, surprise, transformation, and determination. This client's brother committed suicide, and the Moon-Neptune in the 12th related to the fact that she visited his grave every week, which in her Western culture was rather unusual. The sad weekly communion with her dead brother was the manifestation of the resource in question, and it was no doubt an important, empowering experience for her. But, in the consultation, it would not normally be a rewarding exercise to access this kind of energy in terms of resources. This is partly because so many negative memories are associated with it (and the risk of disempowerment is therefore high) and partly because both the Moon in its fall and in the 12th, and the Mars-Uranus-Pluto should be dealt with therapeutically first.

Planetary Influences on the Senses

The emphasis on elements can be strongly influenced by the planets involved. Mercury and Uranus will tend to give a strong air feel, with ideas, speaking, and hearing dominating. Venus and Saturn tend to give more of an earth feel, with physical sensation prominent. The Moon and Neptune brings water and the emotions to the fore. Mars and Jupiter will always give a strong fire feel. In terms of time, Uranus, Neptune, the Sun, Mars, and Jupiter tend to be associated with the present and future, and the Moon, Pluto, and Saturn tend to orient the person toward the past.

The point here is that the mapping of human senses to the astrological elements is not 100 percent reliable. When planetary influences combine with the elements, you will find the disposition to use a certain sense to have been modified. However, when there are strong combinations of, say, Mars and a fire trine, or Mercury and an air trine, or Saturn and an earth trine, or the Moon and a water trine, you can be sure that the corresponding sensory systems—vision, hearing, touching and feeling, respectively—will be dominating.

The Process—Step by Step

1. Choose the Resource. First, you have to find the planetary configuration you want to work with. You will need to decide from the signs involved which elements are dominating. This will normally be a pure element, or either earth/water combinations or air/fire combinations. Choose the most obvious soft aspect, preferably involving Jupiter or Venus. Take care when working with the water element, and remember: Mars, Saturn, Neptune, and Pluto will not have as happy associations as Sun, Mercury, Moon, Venus, and Jupiter.

2. Identify the Activity. Now describe the aspect and its probable effect. This is a difficult process, and it's unlikely that you'll get it right immediately. Tune in and proactively use words, voice tone, and body and eye movements to access the latent resource in the client. By using the right words, intonation, and body language you'll get the client to associate with the resource quicker. It's essential not to be put off by the client's failure to react to the described resource—you'll get there in the end. Keep suggesting possible activities you would associate with the astrological pattern.

The house position is an important clue because this will precisely show where the resource is expressed. Rather than actually describing the supposed resource or the feeling connected with it, it's far better at first to identify the *activity* associated with the resource, and this you can get from the house position. For example, a teacher with Moon in Gemini in the 3rd sextile Venus in Aries in the 12th took her schoolchildren on *cycling* (Gemini) *expeditions* (Aries) to *nature* (Venus) *reserves* (12th house). That's the activity, and once you see an obvious correspondence, as in the example above, you can relax and begin to access the resource.

3. Anchor the Feeling. When the client remembers an activity associated with a conjunction, trine, or sextile, it is certain that they are beginning to feel a very positive energy. However, if one of the planets in that aspect makes a square or opposition to any other planet, they will be inclined to be side-tracked along the more prominent negative energy. Resolutely avoid being led astray by the client into problems, which hinder the energy flow of the resource. And when you are accessing the positive feelings connected with the resource, make no moral judgments. It's not important what the person does, it's the associated feeling you are after.

The resource energy flow is represented by figure 13 on page 81, showing a trine in fire, which I will analyze at the end of this section. When you identify the manifestation of the energy, you are looking at events and phenomena that make up the body of experience for the individual. These form the base or foundation of the pyramid. The middle of the pyramid shows activity and behavioral predispositions. The top of the pyramid shows pure, undifferentiated energy—pure empowerment. It is this pure energy that predisposes the person to a particular kind of activity, which in turn results in a variety of predictable events. What actually happens is that as you identify the activity of the client, you come to dwell together on all the events and phenomena spawned by this activity. When the client does this, he or she feels strong sensory signals. Then, as the client goes deeper into the sensory signals, he or she is drawn into the originating energy, depicted at the pyramid peak.

So, when you have identified the aspect, then the activity connected with the aspect, and then the resource connected with the activity, you have found the treasure. The temptation here is to be satisfied—yes, it was really great to listen to the father telling tales of adventure—but the process is only just starting. At this point the client should be guided into a deeper and deeper experience of the activity or event with as many of the five senses as you have time to encompass. What can the client see, hear, feel? What scents and tastes are there? Notice how the client's voice, posture, and color changes as each sensory anchor digs in. Keep in tune. Figure 12 on page 75 shows which questions are appropriate. In terms of seeing, questions regarding color, size, distance, speed, and dimension are appropriate. With hearing, it is volume, direction, and harmony that are important. With touch, weight, form, and composition matter. With smell and taste, questions concerning intensity, sweetness, bouquet, and delicacy serve to intensify the sensory experience.

The point of these questions is to evoke a detailed sensory map of a particular experience. It's not so much a matter of *what* the per-

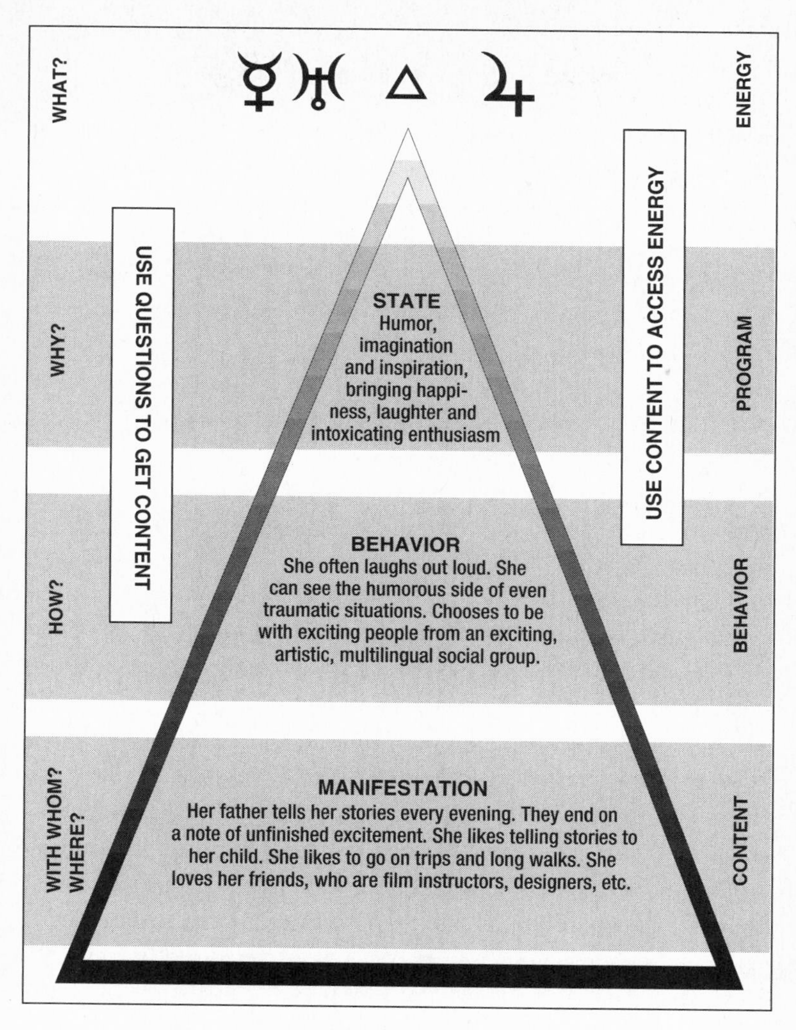

Fig. 13. Dynamics of resourceful energy state. The planetary aspect represents an energy state that gives certain behavior. This manifests as events in the world. When you guide clients to a deep, sensory experience of the events, they access the original energy state.

son feels, but what the sensory signals are. Often the client at this point will begin to use words that evaluate the sensations. "This is a nice feeling," or, "That makes me feel stronger." Don't fall into the trap of evaluations. The questions at this point are, "What is it that tells you the feeling is nice?" or, "Where in your body, specifically, do you sense you are stronger?" Constantly bring the person back to the sensory anchors. These anchors never change—they are always to be found in this specific spot in this specific way. They are unforgettable.

4. Identify Multiple Manifestations. You may wish to identify a string of activities going back in time to the earliest manifestation of the astrological pattern. This serves the purpose of anchoring the person more and more strongly in the feeling of the resource. What actually happens is that when the client starts accessing the undifferentiated energy of the resource, all other manifestations of the resource reveal themselves to memory like pearls on a string. At some point though, you should identify the activity that has the strongest associations and concentrate on extracting sensory detail. At this point, you cease focusing on the multicolored spectrum of manifestations, and you move inward and deeper to access the white light of energy behind the manifestation. It is this pure energy that is empowering. Each planet has its own distinct energy, and trines and sextiles show this energy blending in unique combinations. As you dwell on this energy, it becomes increasingly obvious that this is a unique gift the person has—an amazing blessing. It is not always obvious to the client, though, who often thinks that everyone else has the same kind of gift. It is useful to remind them that this blessing is unique for them. When the cards were dealt at birth, there may have been some jokers, but this is the gift that makes them a winner.

5. Apply the Resource. This process of accessing a resource empowers the client and is useful for pulling the client out of a negative frame of mind. Now that they have been anchored in the resource, they will have a different attitude toward their problems. You can choose to use the resource immediately or save it up for later. You may have been working on a difficult aspect earlier, in which case you can start applying the resource to it, or you can start working on a negative pattern and then apply the resource. How you apply it depends on the individual, but experience shows that an empowered client attacks problems in new, creative ways, quickly finding solutions to them.

6. *Training.* When you have successfully gotten the client to apply the energy of the resource to a difficult situation, it is a good idea to get them to think of two or three areas in which they might encounter similar problems in the future. Run them through these situations, consciously applying the resource. This familiarizes them with the technique so that it becomes an automatic process.

Evoking and Applying Resources: A Practical Example

Hanne, whose chart appears in Figure 14 (see page 84), came to the consultation depressed about the fact that being in love was so painful to her. As soon as she fell in love, she felt a profound anxiety, which then comprehensively sabotaged the relationship over a very short period. This was almost certainly related to the Sun-Pluto in Virgo, in a very loose conjunction with Venus and the North Node squaring Mars. She was born just a couple of days after an eclipse of the Sun, and at an early age, her father had suddenly died. Good-looking and cultured, she was dressed in black from head to foot. Black absorbs light and does not give any light out, and this color choice—without any flash of color to alleviate the impression—is often a sign of being low on resources. It would have been futile to enter into the unfathomable energies of this planetary constellation during the early stages of the consultation. How nice, then, to see Jupiter—strong and in its own sign—in the 3rd house, and trine Mercury-Uranus. It's a fire sign trine, so the energy is in the head, inspirational, and this is emphasized by the fact that both Jupiter and Uranus are involved. Applying the above model, the following section is a short summary of accessing the resource.

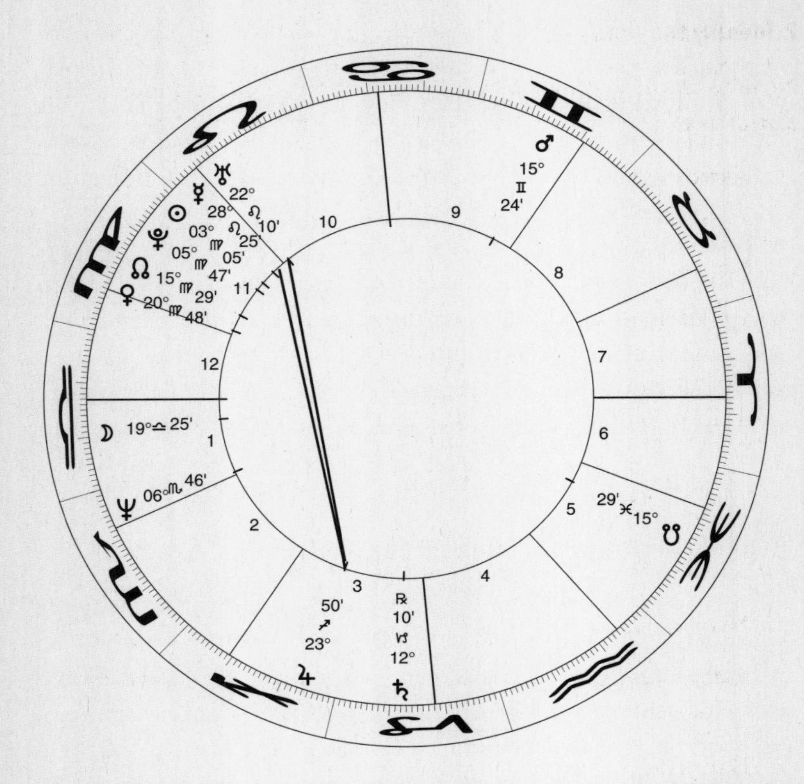

Fig. 14. Hanne S. A trine in fire signs is manifested as a great love of adventure and storytelling.

---

1. Choose the Resource

Looking for a strong planet, and preferably a trine, the Jupiter trine Mercury-Uranus fit the bill. In Hanne's chart, I ignored Mars's trine to the Ascendant; it's much better to have benefics involved, and best with two planets, rather than a planet and a point. Saturn, though strongly placed, makes a weak trine to Pluto, and it's inadvisable to use aspects between outer planets. It's also vaguely trine Venus, but this is not a fun aspect.

2. Identify the Activity

Now it's important to formulate a question that corresponds to the aspect as exactly as possible:

> **Astrologer:** Do you like traveling to exotic countries together with friends?
>
> **Hanne:** Yes.
>
> **Astrologer:** Do you like reading adventure novels?
>
> **Hanne:** Not particularly. [So let's try again.]
>
> **Astrologer:** Do you like modern cinema?
>
> **Hanne:** Oh yes, I do! [Uranus brings modern media into the picture.]
>
> **Astrologer:** Do you do any teaching.
>
> **Hanne:** No, I don't. [OK, so let's try again.]
>
> **Astrologer:** Don't you like recounting adventures to your daughter?
>
> **Hanne:** Hmmm, yes, I love doing that.

After a few questions like this, it turned out that when she was a child, her father used to read adventure stories to her. Furthermore, every evening he would stop the story just before something exciting happened. She could hardly wait to go to bed the next evening to get the next part of the story. So this was the activity most loaded with emotional energy.

3. Anchor the Feeling (see figure 12, page 75)

As the trine is in fire, she made vivid pictures when listening to stories, and these pictures were still crystal clear in her mind. Evoking the images she made at the age of five, she strongly associated with the warm feeling of being read to as a child. As her father's voice was a feature of the activity, it was a simple matter to get her to describe the sound of his voice, and this in turn intensified the sensory experience for her. It's always a good idea to find out what area in the body such intense feelings are connected to, but when asked, she described the energy as something that came down from above. This is a very typical sensation when Uranus is involved; the energy comes down from heaven, so to speak. She sensed the energy of the trine as an aura surrounding her body. Fire trines often manifest as light or energy outside the body.

At this point, Hanne was feeling a powerful energy charge. This was reflected by the color in her cheeks, strongly contrasting with the pale individual who entered earlier—her paleness emphasized by her black clothes. It was particularly noticeable how her voice was imbued with enthusiasm, reminiscent of the child voice urging her father to read on.

4. Identify Multiple Manifestations

If at this point it had seemed that the activity and associated events were not particularly powerful, this would have been the time to cast the net out for variations of the aspect manifestation. Generally, the earlier the memory, the stronger the energy charge, and in this case, Hanne was already enthralled, recalling how happy and stimulating her mental relationship to her father had been. Previously, she had only been able the think of her father in terms of loss—the black hole of her birth Sun-Pluto falling on the eclipse. Now she realized that there were happy memories too.

5. Apply the Resource

At this point, Hanne was radiating—a pure channel for the Uranus-Mercury-Jupiter energy. It was anchored as a sensation of orange, pulsating light around her body. This resourceful state could now be applied to the problem at hand—fear of loss. I did this by getting her on the one hand to feel the positive energy, and on the other to get her to feel the negative energy. When asked what she wanted to do, she proposed enclosing the negative energy with the orange light she sensed from the trine. She performed this process vividly, seeing the dark, negative energy being dispersed by the vibrant, orange light. Inside this container of light—visualized as a glass dome—she felt protected. Normally, this process of juxtaposing a positive and a negative energy would take some time, and it would be necessary to have talked about, and mapped, the way a person's negative feelings are expressed and through which senses. It is also crucial that the client is powerfully feeling the good energy for this to work; if she is not, then there is the danger that the power of the bad energy will overwhelm. It's no use applying a weak resource to a strong trauma, and that's why it's so important to build up a very powerful sensory experience of the resource.

6. Training

After we successfully changed the energy she associated with anxiety, it was easy to run Hanne through a number of situations in which she would encounter the feeling that a person she loved would disappear from her life. We worked on how she would first feel the anxiety and then would sense the glass dome and the positive feeling connected with it.

I had the opportunity of meeting this client about six months later, at which time she was in another relationship that felt much more permanent to her. Her anxiety about losing the person had apparently been completely dispelled. For the first time in her adult life she was entering relationships without fear.

Feel Good Therapy

Working with resources may seem a little complex. The trick is to precisely identify an activity associated with a good aspect, then evoke the senses that are involved. It's so important to believe in the horoscope—the manifestation will be exactly according to the astrology. For example, another client had Venus in Taurus in the 5th, sextile Jupiter in Cancer in the 7th—the two benefics strongly placed by sign. Having emigrated from Romania, she now played the double bass in an orchestra in Copenhagen. The orchestra in the foreign country is represented by Jupiter in Cancer in the 7th, her musical talent by Venus in the 5th.

Jupiter is also the immense instrument, and the sextile shows the hard work as well as the pleasure involved. Having identified the activity, it is simply a matter of getting her to describe what it really *feels* like to play. She put it, unforgettably, like this: "My own pleasure is the most superior feeling." Isn't that great? There really is a hint of Jupiter's exaltation in Cancer with that remark. Accessing this resource was simply a question of her hearing and feeling the double bass playing a wonderful piece of music when she found herself in difficult situations.

It's just wonderful to work with the positive aspects. What I am proposing here is a departure from the exclusive focus on problems and difficulties and a rehabilitation of trines, sextiles, and the benefics as true blessings in the life of the individual. The astrologer actually has the capacity to evoke a strong feeling of happiness in the client. The consultation can be started in this way, and it can be ended in this way, so that the client departs not weighed down by the intolerable burden of his or her negative sides but with a sense of privilege as a bearer of inner magic. If you try it, you may find that it is difficult at first, which is why I have described the process in such detail. The technique is important and worth studying hard. You will find that instead of weakly describing soft aspects for a couple of minutes, then giving up and going on to the hard stuff, you will need at least 10 to 20 minutes to go through this process successfully. You cannot do it any quicker. When you have seen the results of evoking resources, you will never underestimate the power of soft aspects again.

PART TWO

The Major Players—Planetary Combinations

Introduction

The Inner and Outer Planets

For want of better terminology, in the following section I have called the planets beyond Mars outer planets and the planets from Mars in toward the Sun inner planets. I appreciate that there could be a misunderstanding here, as traditionally Uranus, Neptune, and Pluto have often been referred to as outer planets and Mercury and Venus as inner planets. The reason for this distinction is that the Sun, the Moon, Mercury, Venus, and Mars do relate to inner drives and are basic building blocks of personal identity. These bodies are clustered extremely close to each other from an astronomical point of view. At a vast distance from the Earth another type of planet is found. These planets have their own moon systems and are incomparably huge compared to the inner planets. Jupiter, Saturn, Uranus, and Neptune are enormous, gaseous bodies, whereas Mercury, Venus, Earth, and Mars are incredibly dense spheres of rock and metal. While all aspects are important, the combinations of inner and outer planets create easily identifiable behavioral patterns, which I shall discuss in this section.

Pluto is an exception to this categorization, as it is even smaller than Mercury and is not a gas planet. Nevertheless, it certainly does not represent inner personal drives, and it is traditionally associated with deep, collective influences, just as Uranus and Neptune are. I have not written any interpretations for the influence of Chiron in this section. This is not because I do not believe it has an influence—certainly in my own horoscope it seems to have an extraordinary effect. For me, Chiron influences are still under review, and my experience is limited. When dealing with a client in a single consultation I have always found it best to concentrate on the major players and the major aspects, rather than going down the road of diversity. I strongly believe that everything works—from the asteroids to harmonics—but my efforts in the consultation go toward piercing through the background noise and isolating the dominant energies.

Before going on to therapeutic interventions with the difficult aspects, a clear overview of the manifestation of these aspects is crucial. It is essential to have a thorough knowledge of the interplay between the five inner planets—the Sun, the Moon, Mercury, Venus, and Mars—and the five outer planets—Jupiter, Saturn, Uranus, Neptune, and Pluto. These 25 major aspects form a behavioral foundation with specific actions and experiences. Aspects between the inner planets alone or between outer planets alone are naturally also important, but mastering the interpretation of inner to outer planet combinations forms the backbone of any astrological analysis.

The inner planets reflect personal drives and needs that constitute all-pervading tones of being. Their influence permeates every moment of conscious and unconscious life. These tones of being will be modified by whichever sign the planet happens to be in, and the crystallization of these modified energies will manifest very precisely indeed in the house position of the planet. In most cases, the inner, personal planets will also be affected by one or more of the outer planets, and this will result in a specific kind of need or drive, which will lead to specific behavioral patterns, which then will become manifest in the environment. The job of the astrologer is to make the client conscious of the manifestation, illustrate how it relates to behavior, tune in to the basic energy drive, and insure adjustment of behavior to create optimal results.

Many excellent textbooks have analyzed and explained the effects of aspects, and rather than go over the same material, this section will examine them from a practical, therapeutic viewpoint.[8] The assumption is that each of the 25 main aspects will have a negative manifestation. The following descriptions are solely geared to people who actually experience problems in connection with these aspects. There will be a number of people for whom this material does not strongly apply because they manage to channel the energy of these aspects positively. However, for a significant number of clients it will not be difficult to identify the negative manifestation of the aspect, and it may be why they have come. The following descriptions therefore are geared to helping the practitioner to: identify the manifestation of the aspect; identify the behavior responsible and the reasons for it; and suggest alternative behavior based on conscious choice.

Note: I use the Placidus house system in my work, though I may also check Koch house placements. Regarding orbs: the closer they are, the stronger their effect. Clients themselves reveal, through their behavior and their life, whether or not aspects are "within orb."

The Sun—Existential Challenges 4

Any aspect involving the Sun affects the core of our being. Hard aspects will manifest as identity crises at certain times in life, and have a profound effect on how we project ourselves at all times. Dealing therapeutically with difficult Sun aspects is perhaps the most challenging of all tasks, as change has to occur at a spiritual or core level.

Tracing the root causes of the identity nature will almost always bring the astrologer back to the client's relationship with the father or influential childhood mentors. A description of the Sun and its aspects will normally constitute a remarkably accurate picture of the father. There are two mutually inclusive reasons as to why this is so. First of all, it is likely that the individual incarnates in the perfect environment for future development and therefore gravitates to the kind of father best suited for this development. Probably, too, there is unfinished business between them from a karmic point of view, and childhood with this particular father is what both need to make progress. Second, the child will evoke qualities in the father that might not have manifested had the child not been there. Taking the Sun-Jupiter example below, the child will be seeking a mentor and guide in the father, and will ask many questions about the world, evoking in the father an apparently expansive and wise character. The father may not really be like this, but the child will think it is so. In all probability however, both scenarios are likely to be true; the father is wise, and the child evokes wisdom in him.

These distinctions are crucial because many clients complain bitterly about their parents and lay considerable blame on them. But the fact is, parents love their children unconditionally. They may mess up the upbringing, they may behave immaturely, they may do terrible things, they may even dislike their child, but there is a love, biological in intensity, which means that they would give their life for their child. The child is as responsible for the relationship as is the parent, if not behaviorally then certainly karmically. Only when blame is eliminated and personal responsibility shouldered can a relationship change for the better. There is nothing as healing as the reconciliation of a parent with a child, and with the correct view, there really are few impediments to this happening.

Difficult aspects to the Sun from the outer planets show the existential filter through which the child experiences the father. The Sun represents individuation, and the father is the first symbol of this. Bound to the mother's breast and the rhythms of feeding, the child experiences the father bursting in to its environment and disappearing again. In the traditional family, this happens constantly, and each time it does the atmosphere changes for good or ill, the feeling of the mother fluctuates, the earth trembles. There is a sense of disturbance and power. Who or what is this being with such great influence? Some people were brought up in an untraditional relationship, perhaps where the father mothered or even in a lesbian relationship, but the important thing is that, for the child, everything in childhood has profound consequences in the present day. The key is to trace the consequences to the childhood root and show how the same energy pervades the past and the present.

The father is the hero, the light of the child's life, and childhood is the long journey of realization that this god is a flawed human being. The process of individuation is the replacement of the father

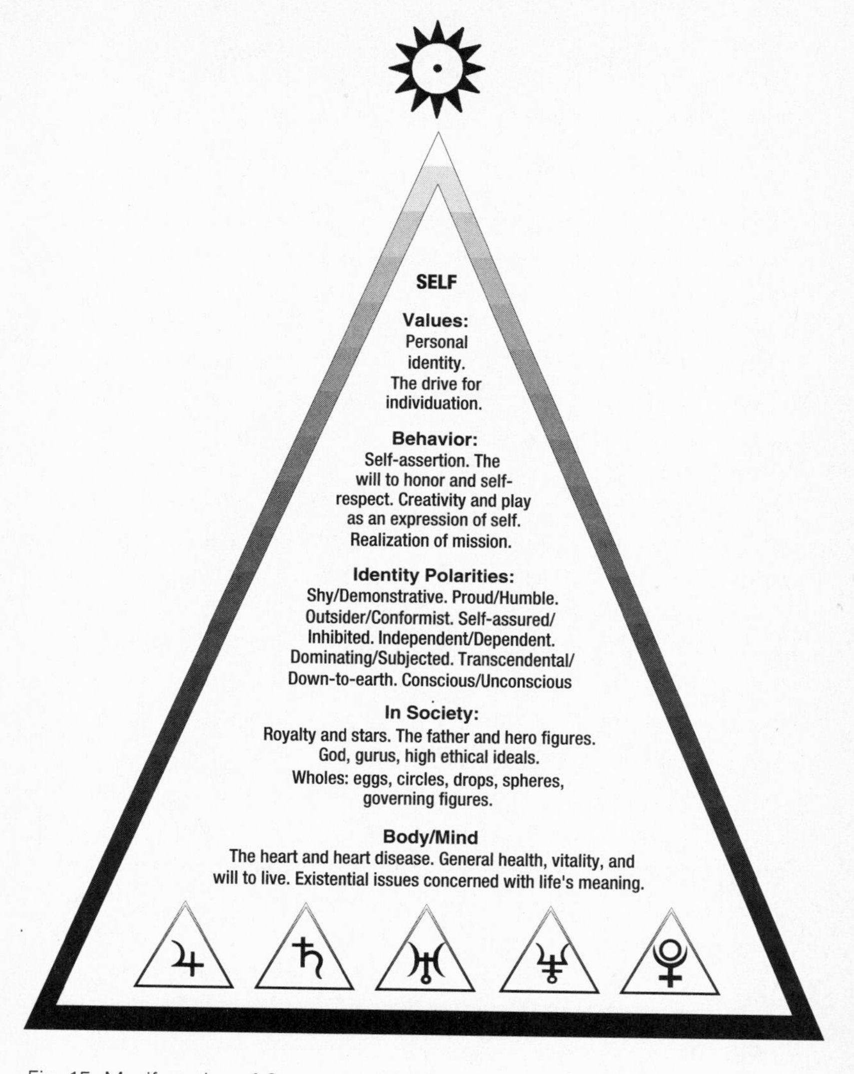

Fig. 15. Manifestation of Sun energy. The Sun's energy, combined with its sign position and any aspects, gives specific behavioral patterns. The most dominating patterns will be shown when the five outer planets are involved.

as the light of one's life with an inner, independent sense of strength and identity. The Sun and its aspects graphically show what barriers there are to this process. This experience is subtly different for sons and daughters. For the son the father is a hero, but a hero to be overthrown in the teenage years. The daughter loves her father with a passion. For a young girl from, say, two to seven, the father is a knight in shining armor. The actions of the father in early childhood have a profound influence on all later relationships. Should the father leave home in this period through separation or divorce, the consequences are far-reaching. Overcoming them will be a major challenge later in life, and this struggle will be clearly shown in the horoscope. The nature of the separation is important. Did the daughter visit the father regularly, or not? (*Separations need not hurt too much. We can leave each other and just be friends*). Did the father cut off all connections? (*My partner will leave me and never return.*) Did a new stepfather enter the home? (*I shall have two men in my life.*) Did the father leave the mother for another woman? (*How foolish, then, to be the wife—much better the other woman.*)

The following descriptions of aspects to the Sun are necessarily generalizations, yet each aspect has a unique identifiable tone that *will* characterize the client's identity. There are hundreds of permutations, but the core energy is the same, and it is through this core energy that change can be wrought.

Sun-Jupiter: The Importance of Being Right

This aspect is an interesting challenge for the astrologer, mainly because the client strongly identifies with his or her own interpretation of the "meaning of life" and is convinced it is right. These types feel they have the answers themselves and perceive everything you say through the filter of their own understanding. This kind of attitude is also very strong with Mars-Jupiter types and to a lesser extent with Mercury-Jupiter. They are smart, and other people may not be so. The client will be looking for evidence that the astrologer is not as clever as reputation suggests, and it's an odds-on bet that the astrologer will at some point reveal the lack of insight and understanding that the client was expecting. This is a key moment for a therapeutic intervention. One client with this aspect interjected, "Yes, but . . ." after everything I said. Translated, this meant: "OK, that was your version of events, but now listen to this, the correct version." My intervention was simply to play it straight:

> **Astrologer**: Do you realize that every time I try to explain something to you, you interrupt with "Yes, but . . ."? You see, this makes me feel like I have nothing to offer you.
>
> **Client**: Yes, but . . . [Which he said yet again, completely unconsciously. He was silent, and we looked at each other as understanding dawned.]

Quite apart from this, Sun-Jupiter people are probably quite learned and have studied in depth to enhance their understanding. You may well be in a situation where you feel outgunned intellectually, and this could awaken any inferiority complexes that you may have. Bear in mind, exactly at this point, that this is standard technique for the Sun-Jupiter person. It's hard for them to resist the temptation to feel superior, and people who surround them tend to go around feeling inferior. This can be a central behavioral habit for them, and the manifestation in the environment is that people around them become sycophantic and do not or dare not give them the intellectual sparring that they need to develop. As this is the last thing Sun-Jupiter types actually want, it can be comparatively simple to get them to see the virtue of a modicum of intellectual humility.

If your Sun-Jupiter client *is* "wise" and superior, it is crucial to lay out your cards and explain exactly how this makes you feel. He has come to consult you but seems to have all the answers himself. What can you do for him? There must be something he hasn't cracked, otherwise he wouldn't be there. What is it, and what is it about him that makes it difficult for him to solve? In cases where the person in question seems intellectually superior, it's far better to enroll his intelligence in the problem-solving process.

There is a deeper existential problem with Sun-Jupiter, particularly with the conjunction. As Jupiter does represent deep spiritual understanding—a meeting with God—and the Sun is the center of being, there is an extremely strong sense of space or boundlessness, which can create considerable anxiety. Sometimes this results in what could be termed "manic" behavior, as if the identity overflows its boundaries. It is the classic signature for hubris. If the client has no spiritual or philosophical structures to contain the insights and experiences, then there can be a sense of being overwhelmed. A strong dose of Saturn in the form of some rigorous mental/spiritual discipline could be of some help. Belief systems have been developed to deal with precisely this kind of challenge. Actually, Sun-Jupiter types are always seeking a mentor to trust and learn from, and many find one at times of spiritual crisis.

Sometimes Sun-Jupiter people feel they're at the center of a deep, spiritual battle, and it is no exaggeration to say that those with the opposition can even pit themselves against, or demand something of, "God." These people may feel that God demands too much of them or is constantly exposing them. Sometimes this is projected more at "the powers that be" in society. Conflict with moral authorities in society can be a constant source of discomfort, both with the opposition aspect, and especially so with the square. These people can go about with a chip on their shoulder and an inner dialogue in which they convince themselves that the moral authorities in their society are stupid. However, in this inner denunciation process, they run the risk of being as obnoxious and dogmatic as the authorities they criticize. Such authorities can be the church, the political system, and influential people in education and media, for example.

Generally speaking, the father is influential, with a far-ranging mind, strongly disposed to venting his opinions about a wide range of matters. There are often strong moral overtones with Jupiter, and the child can hardly avoid having the sense of being judged. Depending on sign position and other aspects, that judgment can sometimes be severe, even though Jupiter also manifests with tolerance and understanding. Because the child also has very strong opinions (testified to by the Sun-Jupiter aspect), intellectual clashes with the father or mentor arise. At a young age the child is put firmly in place, and he or she forms a perception that people in authority are intellectually or morally dominating. They lay down the law . . . and it's so unjust! An inner dialogue develops in which the child becomes convinced of the rectitude of his or her own judgments, conclusions, and beliefs.

In this way, these people come to sift experience through an intellectual, spiritual, and moral filter that subtly changes reality. They see people who have attained positions of authority in the outside world as flawed, and they have an inner self-justification process, often characterized by negative judgments of others. But they *do* have wisdom, and often a simple description of this general scenario is enough to enable this type to modify behavior and create a better social environment. To get anywhere with this type of client it is a good move to defer to their insights and avoid appearing as if you have the answer. They need to be enrolled in the therapeutic process.

What's cathartic to confess for Sun-Jupiter people is that secretly they know they're right and believe themselves to be superior.

Sun-Saturn: The Need for Respect

People with aspects between the Sun and Saturn can often be quite successful in the professional sphere, partly because the drive to prove themselves is strong. However, they are often motivated more by fear of failure than belief in success. Even when they achieve their goals, it is difficult for them to feel satisfied. They long for recognition, yet paradoxically reject it when it is given. Generally, these types will exert a powerful self-control, or struggle to do so. Although they can on occasion exude confidence, especially in business situations, inwardly they feel pitted against the world and in danger of losing status.

Tracing the causes of low self-confidence and lack of self-esteem will almost certainly lead back to the relationship with the father. The perception is that the father was not warm and did not give the client recognition or praise. This type invariably seems to be able to recall an occasion when, coming home proudly with school exam results, the father stares at him or her dispassionately and says, "Well, couldn't you have done better?" The interpretation becomes: "Nothing I do is good enough for my father." It's possible, even likely, that the father was severe and demanding, but the point is that the child is simply waiting for confirmation that he or she is not good enough. This person has a special filter attuned to being "slighted," and he or she manages to interpret a lot of innocent behavior from others as derogatory.

One way of dealing with this loaded perceptual filter is to ask the client to imagine alternative interpretations and motivations for the other person's behavior. Could the father have any other motivation for asking if the child could have done better? The client will look at you blankly, as if no other interpretation were possible. You can ask the client to just pretend . . . *imagine* there was another motivation. With great difficulty, the client will bring up some other clumsy interpretation. Actually, in this example, the motivation of the father is probably quite simple. It is almost certainly *not* to make the child feel small; mistakenly or not, the father thinks that too much encouragement may make the child complacent. The rest is in the mind of the child.

The Sun-Saturn person projects severity onto those in authority and suffers from it. But the implacable judge is within. In the relationship with God—the ultimate projection—God is the Yahweh of the Old Testament, demanding self-discipline, untiring work, self-abnegation, and is not satisfied even when these things have been delivered. It is as if love is lost from the spiritual equation. On a core level of being, there is little self-love or self-appreciation, and this attitude impinges on the environment, evoking events that confirm the inner picture. People who praise the Sun-Saturn person rarely do so more than a few times because they sense that praise is dismissed. Those who give applause require an appreciative response, or they will cease to give it. You have to be able to receive love to give it. Therefore, it is essential that this person make a conscious effort to acknowledge praise and reciprocate, even if the effort is fumbling at first.

Inasmuch as a woman may have difficulty integrating her Sun aspects—as a man may have difficulty with his Moon—the Sun-Saturn is often projected onto men, leading to an attraction to authority figures who create quite restrictive structures around the woman. It is amazing how many women with the opposition aspect have relationships with considerably older men who seem to be father figures. Certainly with both the opposition and the square there is a tendency to deny the inner discipline that Saturn demands and project it on others. These others then have the thankless task of trying to get the person to live up to his or her responsibilities (and then get blamed for being authoritarian). Of course, there is nothing wrong with a relationship with an older person. However, if the client is actually blaming that person for being dominating, or imposing restrictions, then a change of attitude is advisable.

Dealing with this aspect on an energy level means getting the client to go within to contact that leaden voice of authority or the heavy, unloved heart—the inner judge who prefers chastisement to encouragement. This inner judge may believe it is more effective to goad the client into self-improvement, but you can show how ineffective this technique is, and Saturn, after all, does like effectiveness. A good antidote to a strong Saturn is a strong Jupiter, and ideally they should work hand in hand. If resources connected with Jupiter have been activated prior to Saturn—and that means accessing hope, vision, conviction, and belief—then Saturn can be enrolled in more constructive behavior. Results are always more solid when there is more carrot than stick, because results based on fear of failure create a negative dynamic that sabotages success. There will be an inner picture or inner dialogue or inner feeling that Sun-Saturn tries to escape, and in this way failure always looms around the next corner. By focusing the attention of the client on the inner judge, vocalizing the inner critic, and accessing the deep-rooted body sensations associated with Saturn, it is possible to harness Saturn itself to inaugurate a strategy for change.

What works for Sun-Saturn people is to assume a role of responsibility and achieve long-term goals through grit and determination. What's cathartic for these people to admit is that secretly they are convinced they are inferior and that the structures for excellence they build around them will be seen to be a sham.

Sun-Uranus: Visiting Planet Earth

It is always tricky to deal with Uranus therapeutically because this planet is strongly associated with rationality and the mind. First of all, when confronted by those turgid emotions, individuals with this planetary combination have the ability to suddenly dissociate themselves from the matter at hand, looking at things through the wrong end of the telescope. They tend to zoom out and view things dispassionately, as if they were from outer space (and some complain that they feel as if they are!). Curiously, a number of people with strong Aquarian or Uranian qualities, or with a strong emphasis in air signs, can get neck problems because they habitually hold their heads slightly back and at an angle, putting more distance between themselves and the object of their attention than most people. Try it, and you'll get the idea.

Second, these people set great store by being aware, and they may have quite a clear "understanding" of their problems. The astrologer may not be telling them anything new when explaining their planetary dynamics, and it is quite likely that they will evince great interest in the mechanisms involved. The astrologer gets lulled into believing that all is well, yet the consultation can seem curiously flat. The client expertly leads the astrologer on a long detour around an emotional core that does not even seem to be there. Sometimes they can relate quite appalling childhood events, by which the client seems curiously unaffected. In fact, the violently dissociative defense methods of Sun-Uranus mean that emotions are anaesthetized and a strong analytical nature is developed at the expense of emotional vulnerability.

One of the characteristics of Sun-Uranus aspects—particularly the opposition and the square—is a sense of being an outsider in society. In certain cases this can lead to extreme social dysfunctionality. It is rewarding for the astrologer to focus on this area, as it is the result of a defensive pattern in which the individual is trying to protect the integrity of personal identity but is overdoing it. Social alienation is the result of the long habit of dissociation. To deal with this, it helps to find out the original "cause," which in most cases will relate to events concerning the father.

There are several possibilities. It is unlikely that there was much real intimacy with the father, though likely that there was a lot of excitement connected with him. In some cases, the father had a friendly relationship with the child—buddies, even—but on close examination you may find that this friendship was not strong enough to prevent the father from disappearing or being distant at other times. The child gets used to not knowing whether the father is going to be there or whether he will be friendly or remote. As anything could happen, emotional disassociation takes place. You can't flaunt your emotions when the other ignores them completely.

Another scenario is that the father does not accept the usual family projections and makes a point of refusing to be an authority or acts completely contrary to conventional expectations. The child is trained to not look up to hero figures and to not accept or trust conventional models. Already at this stage there can be the perception that the child is up against an unfriendly society. Traditional discipline or paternal guidance is missing. It may also be the case that the father cannot exert any kind of traditional authority anyway because the Sun-Uranus child is a rebel and just makes too much trouble when disciplined. It is well-known that the way to get a strongly Uranian child to do what you want is to order the child to do the opposite.

The nature of Uranus is the quality of surprise and the unexpected. Therefore the painful memories of the past are often associated with sudden events that cauterize the emotions, rather than a long process. The sudden disappearance of the father, an accident, a period spent abroad, sudden domestic disruptions—all these things can provoke reactions that create behavioral patterns that can be at the root of a troubled environment later in life. When dealing with these issues it will slowly dawn on the astrologer that the cool and dispassionate style of Sun-Uranus is a means of coping with events that otherwise could be very disturbing. Familiarizing the client with these issues may trigger a sudden transition from rational detachment to being emotionally overwhelmed, and the astrologer should be prepared for this.

Women with this aspect will often reject traditional relationships, and they are certainly not comfortable with men who imagine themselves masterful and in control. Quite a number will choose partners from untraditional cultures (like, for example, Islamic), both because of an attraction to the novel and challenging and as a social statement. Paradoxically, it is often the case that men from these cultures actually expect a certain amount of subservience or obedience, which leads to extremely interesting tensions. I have encountered many a puzzled Iranian, or Indian, or Israeli who cannot figure out the complex and contradictory needs of their Sun-Uranus European partner.

What works for Sun-Uranus people is to create an independent role for themselves in society and to be an example to others through manifesting unique, modern, and original qualities. What's cathartic for these people to admit is that secretly they know they are outsiders and that those they know will discover how distant they feel.

Sun-Neptune: Am I Real?

As the Sun represents the inner core of identity, and Neptune is a transcendent influence whose primary effect is to dissolve attachment to the concrete, the effect of this combination is a sense of unreality about personal identity. Whereas many people never get around to questioning who they are, and why they are on Earth, existential questions arise inevitably and naturally with this aspect. It is difficult for this type to hang on to a sense of personal pride or self-esteem because the stronger the experience of identity the greater the sensation that it is, somehow, unreal.

The Sun-Neptune client has the magical ability to be almost anything or anyone he or she chooses to be—the trick is to make a choice that is going to be satisfying in the long run. Everything is imbued with a sense of impermanence, as if the client already realizes that anything built up will also disappear in the sands of time. Why bother, then? Generally, this core awareness makes it difficult to assert one's ego because of an innate sense that the foundations of identity are an illusion. Perhaps the client is on to something here, realizing basic spiritual issues the majority of people never allow themselves to confront. Buddhism, for example, teaches that the sense of identity arises from a constant process of self-creation, moment by moment, whereby the forces of a person's karma create an ongoing energy. This is the energy that constitutes the self, and it is upon this constituted self that our character and lives are built. Deeper and deeper meditation would bring the practitioner further and further back into a state of being (or nonbeing) prior to the constituted self, and, indeed, this is generally taught to be the process we all go through when we die.

It is extremely helpful for the Sun-Neptune client to have a spiritual perspective to handle the sense that life is empty and meaningless. Life may be so, but there are a lot of people around, and reality for many of these people is struggle and suffering. Meaning for the client can be found in a profound sense of identification with, and compassion for, people. This can lead them into career areas where social consciousness plays an important role. Many people with this influence turn within at some stage of their life, using some spiritual practice like yoga or meditation to experience more and more subtle states of being. It is possible for Sun-Neptune to get a very strong sense of self in this way—and a strong sense of self is a prerequisite for having power in life.

The experience of the father figure through the filter of a Sun-Neptune aspect is often one of powerful idealization. Even with the square or opposition there is a strong sense of identification or even merging with the father, and a feeling of truly understanding his deepest nature. It is possible that the father reacts to the way he is idealized by the Sun-Neptune child by revealing more of the beautiful sides of his nature and by initiating the child into his own inner world. The child tends to filter out aspects of the father's character or behavior that do not conform with the ideal at an early stage, just as the father instinctively tends to project those qualities that please the child. While idealizing the father, the child may form the perception that the father is the victim of the outer world or misunderstood in some way. It is usual for father and child to collude in blaming the mother for unfeeling behavior.

Naturally, this idyll cannot last, but it is often only in early adulthood that the client realizes that the image of the father is *not* the father, and that he is/was just a human being with both good and bad qualities like everyone else. Indeed it may be discovered that the father misrepresented the world because of his own character weaknesses—with all the consequences this now has for the client's world—and it can be a major transformation when this realization dawns. It is naturally important for the astrologer to ascertain whether this idealization of the father has taken place and whether the client is clear about this projection and ready to take responsibility for it.

If the female client has not become aware of this projection mechanism, then in all likelihood she will choose a partner who embodies her ideals but experiences himself as misunderstood or frustrated by the world. This may be so, but inasmuch as the client herself is not able to realize her own dreams and visions, they will be projected upon the partner, who will then be experienced as someone not able to bring *his* visions into reality. A lifetime of experience has given the client high ideals but low expectations, and this often translates as a sense of resignation (especially if Saturn configures with the Sun-Neptune aspect) or cynicism (if Mars configures).

In the last analysis, the world without and the consciousness within are two very different things: the world becomes what it is because of the planetary filters through which it is experienced. Sun-Neptune people are more equipped than most to understand this, and therein lies the antidote to the existential doubt that they can experience. These people tend to get daunted because the world is not as they wish. Often, Sun-Neptune people will focus on things in the outer world that exacerbate a sense of disappointment, whereas what they need to focus on is what is practicable for them. It may be that all human endeavor motivated by ego dissolves, but endeavor motivated by ideals of a higher order can stand as a beacon for humanity into the future.

What works for Sun-Neptune people is to aim for spiritual insight and to get involved helping others who are suffering in one way or another. What's cathartic for these people to admit is that the individual they project themselves to be is empty and meaningless and that people will see through their act.

Sun-Pluto: Anxiety and Dominance

People with Sun-Pluto contacts come across very powerfully. Often they possess unusual and magical skills, and those who have managed to transform and integrate the influence of Pluto can be very impressive, merging considerable psychological skill with great charisma. However, as with all Pluto aspects, there is anxiety about not being in control, including during the consultation. Sun-Pluto individuals tend to project themselves very strongly onto their environment. The effect of this is that others experience them as being overwhelming. Anxious that others may dominate them, Sun-Pluto individuals compensate and end up being dominating themselves, often without being aware of it. They are concerned about how to survive in any given situation and often oblivious to their effect on others.

There seems to be a vacuum at the core of identity, which is often quite literally experienced as a black hole around the heart area that sucks out the life force. It's as if Sun-Pluto individuals have to invest twice as much energy in simply being themselves. In extreme cases, they fear becoming invisible, empty, or annihilated. This fear probably originates from near-death experiences in the womb or early childhood. It may be impossible to find the root cause, but not so difficult to find a series of events during the client's life that they have perceived as life threatening. The consequences are that these individuals interpret events in their environment as more threatening to their identity than they really are, and they overreact, thereby creating a crisis atmosphere that actually conjures up power battles.

Central to this kind of behavior is the client's perception of the father's influence during upbringing. Generally, it is as if the father were not present. He may have literally disappeared, or if present, it appears that he ignored the child. There is the perception that the father is hidden behind the pages of a newspaper, to emerge at intervals with the sole purpose of trying to mold the child to his will. Sometimes there is the sense that the father was not the biological father (and he may not have been); Pluto has a kind of ersatz effect, replacing the natural with the unnatural. This sense of alienation that the child feels from birth is a major thread that runs all the way through life. Rejection of the father has spiritual consequences, which makes it very difficult for the person to make contact with the foundations of the Self.

It is of course possible that the father did try to make contact with the child but had difficulty doing so because of the child's very nature, and this is a good initial therapeutic approach. It is as if the child is just waiting for the opportunity to amputate the connection to the father, and any excuse will do. At some point, there will be a key misinterpretation of the father's actions, which will be used to justify subsequent rejection. It is extremely useful to identify this moment in childhood. It is almost certain that the Sun-Pluto child is driven to overthrow the authority of the father from day one. This in turn may evoke extreme behavior, ultimatums, and battles of will, which can only end in the child's humiliation. It's a no-win situation for the father, too: if he allows himself to be intimidated by the child, then the child's perception is that the father failed through being weak, and this is a very common manifestation of Sun-Pluto. If the father insists on maintaining his authority, then childhood is a long battle with both child and father as losers.

The astrologer will probably experience the dichotomy of power and powerlessness from the moment the client enters. Sun-Pluto people have no respect for the astrologer who waffles through the consultation or allows them to take control, which they may well try to do. At some point, the astrologer must replay the father/child scenario described above. The astrologer has the opportunity to rip away the veil that affects all the client's relationships, which is that the power struggle with the father repeats itself because at the root of being there is a deep existential anxiety. At the moment when the client wrong-foots the astrologer and the balance of power in the astrologer/client relationship slips, this is the time to draw parallels with what happened with the father in childhood. Sun-Pluto is a very grateful aspect because it contains the seeds of transformation in an innate, psychological wisdom. When the client learns that the consequences of existential anxiety lead to power games that make other people uneasy, they can spontaneously transform. The focus needs to change from concern about personal survival to the awareness of the dilemmas of others.

Women with Sun-Pluto aspects may find themselves projecting their powerful shadow on male partners and authority figures. The power and authority that Sun-Pluto suggests in the woman's character may not be apparent at first, but it *is* there. At its root is the perception of the father as a remote and autocratic figure, or as weak and seriously flawed, sometimes psychologically. This type is attracted to secretive men with charisma and power—men, perhaps, who are able to protect her from the dangers of the world. But, as the Sun-Pluto aspect also implies willfulness and deep-seated power, the chosen man will find his power and authority undermined in the long term, and a sense of powerlessness and even desperation creeps in. Overly concerned about any apparent weakness in the male, the woman probes and examines his personal flaws, and they magically open up to become yawning chasms for the relationship to crash into. Everyone has flaws, but successful relationships concentrate on what's good in the relationship, while giving support to what is weak. Sun-Pluto tends to focus on what is structurally weak and then gnaws away at the weakness, precipitating the disappearance that is so feared. This person needs to know that her own behavior can be responsible for the gradual weakening of the male and her ensuing loss of respect for him and that this is a childhood replay. Then the person needs to confront the anxiety behind this behavior, own it, and allow it to transform.

In the preceding analysis, the themes of anxiety, power and powerlessness, control and will relate to a sense of emptiness or void at the center of being, which the client constantly skirts, never actually confronting the demon within. Using sensory techniques described in this book—once early issues have been identified and related to the present-day issues that are their consequences—the astrologer can guide the client into a deep, sensory experience of the actual Sun-Pluto energy. A black hole has actually sucked light into itself for eons and, as such, is a reservoir of awesome, transformative energy. It is possible for the darkness to turn to light, and for the client to completely turn around the perception of self and be reborn as a powerful and conscious individual.

What works for Sun-Pluto people is to embrace positions of power while working on their own psychological development. What's cathartic for Sun-Pluto people to admit is that they feel powerless and are frightened of losing control.

The Moon—Cradle of the Unconscious 5

In Western astrology the Sun plays an exalted role, possibly because the individual and the process of individuation are a major driving force. In the East, on the other hand, the Moon plays a more important role in astrology, reflecting more the culture of family, ancestors, and belonging. Seen from the Earth, the Sun and Moon are equal in size, and as far as the psyche is concerned, they are equally powerful in the horoscope. The power of the Moon is awesome. Studies show that only a minute percentage of our daily life is lived consciously, the rest is habit; we live like automatons, repeating most actions without any conscious thought. We can perform a complex series of actions, such as driving to work, and yet have no conscious memory of how we did it. Aspects to the Moon show what forces shape our unconscious, thereby defining our automatic reactions to stimuli in the environment. Hard aspects will show what defensive patterns we build up to create a sense of security and comfort, and these patterns have a profound effect on how our personal environment coalesces: who we surround ourselves with, the house we have, the place where we live—everything.

If we remained unconscious, we would never be able to transform any of the influences from the Moon. Sooner or later—and generally around the first progressed Moon return and Saturn return (from 27 to 30)—we find that the defensive structures we create also imprison us, and at this point we can consciously choose to shine the light of awareness on these unconscious patterns.

Aspects to the Moon will always relate to how we were nurtured and the environment we were nurtured in, and generally this means the mother. From the womb to the breast and on into childhood, the mother represents pure love and a haven from the storms of life. The Japanese character for love is an ideogram representing mother and child. The mother loves the child and would give her life for it, and the child loves the mother, feeling totally merged with her. However this is not always the story the astrologer hears from clients, who often complain bitterly about their mother. Complaints and descriptions of the mother will reflect the aspects the Moon receives, and these aspects show how the environment is filtered and warped via the unconscious perceptions of the client. You cannot know whether the mother is as the client describes—she may be, but it hardly matters. The aspects to the Moon may describe karmic influences indicating what type of mother the child will gravitate toward, but they will also describe behavioral patterns that will *evoke* certain qualities in the mother—and, later on in life, in others.

The Moon is like the barometer of the ocean of feelings within us that creates our mood, which in turn affects the atmosphere around us. Ideally, this ocean should be calm and still, and, as such, a perfect mirror for our soul or Sun. In reality, environmental stimuli whip up this ocean into a choppy sea, which obscures clarity, or they create tidal waves, which threaten to overwhelm us. At this point, we react unconsciously to minimize danger, build defensive bulwarks, and create buffer zones. The Moon and its aspects show what kind of behavior we instinctively adopt to protect ourselves, but it also shows what kind of receptive channels we can create to interact successfully with the environment. A man with a Moon-

THE UNCONSCIOUS

Values:
Instincts. Feelings.
The urge to protect.
The ability to be receptive.
The need for security.

Behavior:
Instinctive reactions. Merging and attachment. Development of habits. Vulnerability and protection. Bonds.

Emotional Polarities:
Dependant/Distant. Sensitive/Blocked. Insecure/Secure. Defensive/Open. Vulnerable/Tough. Giving/Entrenched. Calm/Tense.

In Society:
Childhood. The Home. Family. Mother. Milk. Atmosphere in the home. House and land. Ancestry. Bed. Night. Dreams. Water and its magnetism.
The occult. Witches. Mirrors, photos. Silver. Boats. Hotels.

Body/Mind
The stomach and breasts. Body liquids. Left side of body. The womb. The memory. The emotional life. Ingrained habits. Unconscious bonds.

Fig.16. Manifestation of Moon energy. The Moon's energy, combined with its sign position and any aspects, shows emotional needs and unconscious drives, which have been cemented through childhood experiences.

Neptune aspect, for example, may protect himself by withdrawing into a protected or isolated environment, but the same aspect will show an extraordinary sensitivity to fellow human beings.

Just as women tend to project the solar qualities onto male icons, men tend to suppress their emotional sensitivity, evoking in the women with whom they come in contact behavior that reflects their own lunar aspects. This is neither ideal nor wise, but gender roles in Western society insure that this will continue. Men have a macho image to take care of, so they tend to ignore emotions, and at considerable cost. Developing nurturing qualities will make the male more whole, just as developing leadership qualities will make the female more individuated. The astrologer should be aware that when describing the Moon to a male client, the description might appear to fit his female partner better. When personal characteristics are suppressed, they tend to manifest in intimate partners in an aberrant form, so it is crucial that lunar projection mechanisms are explained. Working with aspects to the Moon is very rewarding because it is relatively easy to identify unconscious behavior and then resolve to develop new behavioral patterns.

Moon-Jupiter: Being Unconsciously Judgmental

It has been said that Jupiter will always bring great benefits, even with the worst placements and aspects. The astrologer trying to find deep character flaws with Jupiter-Moon aspects is going to be hard-pressed. From a behavioral point of view, the problems here can be to do with being emotionally overwhelming and perhaps somewhat gushing. This is hardly an area that needs therapeutic attention. More serious can be an inflation of emotional reactions and heavy moral overtones, which are often associated with Jupiter when it negatively aspects a planet. As with all Moon aspects, the formative influences are to be found in the emotional environment of childhood, specifically the feeling tone of the mother. In some cases, then, the mother laid down behavioral laws that strongly shaped the client's later judgments and attitudes. This is particularly true when either planet is involved with the 9th house, where moral judgments become quite a dominating factor.

As Jupiter has a lot to do with core beliefs, the Moon's contacts to this planet show a loyal, instinctive attachment to the beliefs and values of childhood. This can hinder objective discussion of important issues, because the basic energy of Moon-Jupiter is an unconscious faith in tried and trusted values. They *feel* what is right or wrong, with a deep conviction but little logic. On one level, their instincts are extremely good; on another, it can be an exasperating business getting them to feel a degree of emotional objectivity. With the square and the opposition, they will come into conflict with society, because their gut instincts will tell them when something that society judges to be right is in their eyes wrong. They may be right, but the danger is that their feelings will be out of proportion to the issue or that they will get emotionally outraged over something they can do little about. In some cases, the client feels powerfully that an injustice has been done, and this will often relate to perceived discrimination as a child. Therapeutically speaking, it is the emotional filter, which registers "injustice," that needs attention. As with all Moon aspects, behavioral patterns are totally ingrained and habitual, and there has to be a strong motivation to create change.

In nature, the Moon and Jupiter are very different, as the Moon has an instinctive need for protection and security, whereas Jupiter invokes a sense of adventure and freedom. This can sometimes mean that the Moon-Jupiter client has difficulty reconciling domestic pressure with the need to expand his or her horizons. In women, maternal drives will conflict with the intellectual. It's difficult to go to evening classes with a six-month-old baby. As the Moon's emotional drives are often repressed in men and evoked in their female partners, Moon-Jupiter men may send mixed messages to their partners regarding this dichotomy between domestic security and cultural stimulation. But all these things are unlikely to be the stuff of trauma. Generally, Moon-Jupiter aspects show tremendous emotional resources. Problems that do arise come mostly from overindulgence and lack of structure in daily life and from irrational convictions or a strong sense of injustice. Training in objectivity, planning, and discipline would go a long way to correct imbalance here.

What works for Moon-Jupiter in life is a free environment, "home on the range," which is stimulating for the growth of those they love. What's cathartic for Moon-Jupiter people to admit is that they are emotionally demanding but do not want to be held accountable for fulfilling other people's emotional needs.

Moon-Saturn: The Emotional Drawbridge

In the consultation, freeing up the energies locked in the Moon-Saturn aspect is always a much tougher nut to crack than the astrologer might expect. While the issues of emotional hurt with the Moon's hard aspects to Saturn are obvious, years of building defensive armor effectively prevent the astrologer from accessing the client's inner feelings. The impression and aura of Moon-Saturn clients is one of self-sufficiency and maturity, but locked within is a seven- or fifteen-year-old child trying to get out. This reality will not be immediately apparent to the astrologer, but to anyone living on an intimate footing with the client, it will.

As always, the best way of motivating the client to make changes is to identify the behavior associated with an aspect, and then to describe the consequences of this behavior. The client is completely familiar with these consequences and would generally do anything to avoid them. Herein lies the engine for change. The behavioral patterns of Moon-Saturn are associated with a self-protective aloofness and a reluctance to take emotional risks. This can mean many missed chances for having a rewarding emotional relationship. Initially, then, they are surrounded by an isolation based on a reluctance to be open and vulnerable. As the consulting astrologer, you probably will be immediately sensitive to the aura of reserve around the client, and you will note your own inclination to keep a respectful emotional distance. This is the first behavioral stage in the young adult.

However, once intimacy is established, and the person opens the drawbridge, then a great emotional hunger is revealed; the hunger of the seven-year-old who was never allowed to be a child, the hunger of the fifteen-year-old whose teenage life was tied up in duties or family struggle. Often there is clinging behavior: the hug that might normally be expected to last a few seconds continues until the partner is forced to detach. For the person attracted to and now intimately involved with the Moon-Saturn individual, it is quite a surprise to experience this sudden immaturity and vulnerability, and he or she may be ill-equipped to deal with it. A typical reaction at this stage is for the partner to withdraw emotionally, and this in turn confirms the worst fears of the Moon-Saturn person, who has a history of past rejections. Moon-Saturn will then go into defensive mode, resolving never to show emotional vulnerability again in the relationship. This is extremely confusing for the partner, who can become really disoriented by the change from childish sensitivity to middle-age cool.

It is crucial to cut right through this circular behavioral pattern, which in the long run deadens love. Moon-Saturn people must realize that no partner can supply the emotional nourishment that they have always perceived to be lacking. They must work on their own filters, taking off the protective armor that inhibits emotional expression. Having done so, they must dare to be open, and be resilient. In the long run, this aspect gives emotional self-sufficiency, and this is achieved not by making emotional demands on the partner but by being the pillar of support the partner was initially attracted to.

Both the Moon and Saturn are associated with parental influences, and when in conjunction, square, or opposition, it is unlikely that there was much family harmony. However, Moon-Saturn aspects chiefly reflect the relationship with the mother. Generally, the client perceives this relation as difficult, but just as often there is a strong bond and loyalty to the mother. The common factor is, however, that the mother was unable to give unconditional emotional love to the child. Love is conditional on the child's performance. Often this aspect relates to the "grown-up child" syndrome—a childhood in which responsibilities were reversed, with the child feeling a strong responsibility to help the mother for one reason or another. As Saturn has such a strong traditional relationship with the concept of karma, it is extremely likely that there is a mysterious bond between mother and child that involves sacrificing the carefree nature of childhood for a heavy responsibility. However, anyone with a Moon-Saturn aspect is born with a strong motivation to take responsibility. In relation to the mother, this results in an ambivalent attitude. There is a motivation to shoulder responsibility, coupled with resentment at being expected to do so.

It is difficult for any child to take on the mantle of adulthood at an early age. In retrospect, the client will be able to remember any number of situations when childhood expectations were crushed, when hoped-for love became rejection, when duty prevented play. Each of these events were character building, each of them forged another link in the chain mail, which, in the long run, lowered expectation of reward and belief in serendipity. What astrologer and client can aim for—apart from avoiding behavior that perpetuates the negative consequences of being overly defensive—is an appreciation of the wonderful human qualities of a person who has developed self-discipline and who is prepared to take responsibility. Self-appreciation is a great antidote for Saturn aspects.

What works for Moon-Saturn people in life—but does not necessarily help them on an emotional level—is their ability to take on responsibility uncomplainingly and deal with difficult issues patiently. What's cathartic for Moon-Saturn people to admit is that they feel starved of emotional support but are frightened they'll be rejected if they show vulnerability.

Moon-Uranus: The Restless Gypsy

Moon relates to mood and atmosphere, and the aura of Moon-Uranus clients is often one of electricity and magnetism. They create quite an exhilarating environment around them. On top of this,

Uranus has the capacity to bring the light of consciousness to bear on what it touches. Given that the Moon's domain is the unconscious, this is an interesting mix. The astrologer will often find that these clients are very aware of, yet extraordinarily dispassionate about, the emotional issues in their lives. They already have dealt with a number of issues in self-development courses, and generally, they seem at home in group-oriented ventures. As you describe possible childhood influences, they may supplement with helpful comments of their own. You cover the emotional ground quickly.

So what's the problem then? I remember a client, with a strong Moon-Uranus conjunction in Cancer in the 10th, who told me that when she was a young woman, she brought her boyfriend home to meet her mother, who subsequently seduced him. How did the daughter react? Not too drastically—she was still good friends with her mother. As you deal with the emotional issues of difficult Moon-Uranus aspects, you can discern a kind of spaced-out sensation, or numbness. When in crisis, Uranus reacts through a wrenching detachment, as if sucked into space at the speed of light. Moon-Uranus can be likened to an astronaut on a space walk, delicately attached to the mother ship through a silvery life-support system, and in danger of being cut off at any time. What this translates as in real life is a sense of transience and the tendency to flit from place to place without creating bonds of belonging. These people can be very happy in foreign environments that constantly provide new social and cultural stimuli, but they have difficulty cuddling up to a partner in domestic bliss.

A common Moon-Uranus complaint in relationships is that the partner spends too much time in front of the television. As the astrologer, you may be inclined to buy this—yes, people do spend too much time in front of the TV, don't they? But the issue is this: when watching TV, two minds commingle in shared mental and sensory processes, both people see and feel similar things simultaneously. It can be very intimate. But Moon-Uranus does not know how to cuddle on the sofa in front of the tube, and restlessly wants something to happen. Complaints that the partner is wasting time watching TV need to be examined in detail. How much time? If it's two to three hours a night, then it's the national average, and so what? What you are looking for is evidence of restlessness and inability simply to "be" or to relax.

When examining the relationship to the mother, you'll often find that Moon-Uranus people have a "friendly" relationship to the mother. Often the mother is unusual in some way—has alternative interests, comes from a foreign environment, and has "enlightened" views about bringing up the child. The child is brought up as an equal, and the atmosphere is democratic. Behind this, however, can lie a profound unease with the child having any feelings of dependence. The mother shares the ideas but does not wipe away the tears. Of course, it is Moon-Uranus itself that has a specific kind of behavior that evokes a specific reaction in the mother. The Moon-Uranus client may not allow the mother to come in close, may prefer the bottle to the breast and the play school to the home.

The key issue in childhood is the probability that there was a sudden event that radically changed the client's whole attitude toward issues of belonging and security. Seen through other people's eyes, this event need not be particularly traumatic, but the Moon-Uranus filter is almost waiting for a childhood upheaval to latch on to. A sudden removal to the hospital, a separation, a move, a violent event—something happens that initiates an unconscious decision not to believe in safety or security or belonging any more. A cauterization takes place, the wound is sealed, but it resonates into the future, and this accounts for the strange sense of emotional alienation that Moon-Uranus people feel.

As with all Moon aspects, the way in from the point of view of astrological intervention is through the mood or through feeling. The access point for Moon-Uranus clients is their peculiar state of not feeling anything in particular, when they should. Dealing rationally with emotional issues, as explained earlier, is home territory for Moon-Uranus. Here they will outperform the astrologer, but it won't help. However, through the *feeling*, they will access the event, and through the event they may access the decision their Moon-

Uranus aspect predisposed them to. Becoming conscious, they can choose to change, and reap the benefits in the form of increased emotional commitment in their present-day life.

What works for Moon-Uranus clients is an emotional life free of commitments, where they are at liberty to do what they want, when they want. What's cathartic for them to admit is that they feel as if they are on the outside looking in and that a sense of being an outsider sits uneasily in their stomach.

Moon-Neptune: The Psychic Sponge

This combination represents an energy that evokes the greatest sensitivity of all lunar aspects. The protective structures and boundaries that normally envelop the emotions seem absent, and this in turn increases the likelihood of being hurt. Sensitivity to others—particularly with the opposition aspect—means that these individuals go to great lengths to tune in to what others are feeling and to adjust their behavior to minimize bad feeling. They act as an emotional barometer, and react unconsciously and caringly to imbalance and suffering on an individual, social, and environmental level. In personal relations, they can have great difficulty discriminating between their own feelings and those of other people because of a tendency to absorb and take on the emotional state of others. They overidentify and tend to overdo consideration, and this is partly to do with a desire to avoid emotional confrontation.

With this aspect, mood tends to take over and dominate the psychic state. It is as if Moon-Neptune individuals have their own private lake of sorrow, with which they are intensely familiar. When emotional upsets arise, they step into the lake and swim listlessly about, sad to be there but glad to give themselves up to the siren song of a place where they paradoxically feel secure. The temptation to enter this psychic sea is great, but the way forward for Moon-Neptune people is to resolve *not* to immerse themselves but to circumnavigate the lake—to be aware of the maelstrom within and choose not to enter. The benefits are twofold: first, they avoid creating a depressive atmosphere that weighs down people in their immediate environment, and second, they come to realize that the attraction to indulging negative feelings is just a habit, and with practice it can be broken. There are no benefits whatsoever from swimming in the lake.

Naturally, the desire to take mood-enhancing substances that can lift them out of sadness—such as pills, dope, and alcohol—is strong, but this will inevitably exacerbate the whole process, making it more and more difficult to develop emotional clarity. When Moon-Neptune is even partly subject to conscious control, the client can access tremendous insight, intuition, and clarity.

The consulting astrologer will probably discover in Moon-Neptune clients an area where the ship of life leaks, where they are floundering emotionally and profoundly unhappy. Where Neptune manifests, you learn to surrender; it is through the dissolution of whatever structures that have been created that you learn about the transience of human life and gain insight into divinity. Having said this, unhappiness is not a necessary condition of these insights. Material rewards may not arise, but spiritual rewards are the promise of Neptune.

A Moon-Neptune client tends to describe his or her mother as an unhappy person and in all probability something of a martyr. The sympathy with the mother is quite often accompanied by an antipathy to the father, because the child identifies the suffering of the mother with the difficulties she has had in her relationship. Whatever the actual emotional state of the mother, the important point is that the Moon-Neptune filter is finely attuned to any emotional balance. The mother only has to groan or sigh, and that sigh directly enters into the heart and confirms what Moon-Neptune has always known: "My mother is suffering . . . and I am responsible." Mother and child develop mutual dependence: the mother develops behavior that is rewarded by the concern of the child, and the child wins love through solicitousness. Voice tone and other subtle signals become the language through which they communicate, and later in life mature relationships are also trained to develop in this way.

However, since the future partner may not be skilled at this form of communication, misunderstandings arise and feelings are hurt. It helps for Moon-Neptune people to put words to feelings and *verbally* insure they are understood.

It is important to convey to Moon-Neptune clients the connection between their overattunement to the mother and subsequent feelings of sadness, guilt, and ill-defined responsibility. Current sadness in emotional relationships is based on patterns and decisions established very early in childhood. I remember a client who, via primal therapy, actually recalled the moment of birth when the mother screamed and cursed in a state of great pain. At that moment, the client discovered he had felt tremendous guilt at causing his mother such pain, and he resolved to do what was in his power to alleviate her suffering. And he did, from doing housework and washing the dishes to sympathetically colluding with her against the father—her "hard-hearted" husband. The reality was that the mother always made a fuss about pain—it was not a reflection of deep suffering—and the child had resolved to live a life based on a false assessment that stemmed from the energy dynamics of his own Moon-Neptune aspect.

What works for Moon-Neptune clients is a life free of too many disturbances—in peaceful surroundings, in nature, by water—in which they can withdraw into protective isolation. What's cathartic for them to admit is that they use the threat of sadness and the ensuing bad atmosphere to force others to organize the environment in a way that suits them.

Moon-Pluto: Amputating Feelings

This combination gives access to the very deepest states of emotion, resulting in psychological insight and unerring instincts if the underlying anxiety connected with the aspect is brought to the surface and dealt with. As the Moon's greatest need is for security, and Pluto always brings constant transformation, Moon-Pluto clients tend to live with the fear that forces in their environment will take over and render them helpless. As with all Pluto aspects, fear of being powerless gives a sense of emergency that drives the individual to establish control, and with Moon-Pluto this leads to dominating behavior that evokes discomfort in others. During the consultation, if and when the astrologer feels uncomfortable, this often signals the way in to initiate a process of transformation for the client.

The difficulties and traumas connected with hard Moon-Pluto aspects are hermetically sealed and consigned to the dustbin of the unconscious. To bring them up to consciousness not only requires confronting frightening emotions, it also means casting aside the cloak of invulnerability that Moon-Pluto clients appear to wear. It takes a lot of skill to get the client to open up this can of worms. Nevertheless, all behavior has consequences, and if the client becomes aware of these consequences, he or she will want to make a change. The consequences of a badly managed Moon-Pluto aspect is that other people around the Moon-Pluto person feel insecure—in extreme cases even to the point of becoming near-psychotic. This is because Moon-Pluto has the ability to amputate all feeling at times of crisis. And Moon-Pluto tends to evoke crisis. This leads to exhausting emotional relationships and the sense that the intimate life is a battleground or a wasteland.

At times of crisis, Moon-Pluto is solely concerned with survival and has no resources to consider what is happening emotionally for other people. Inside, they have a feeling of emptiness or deadness. While Moon-Neptune, for example, would feel sorrow, Moon-Pluto cannot allow any feelings. Other people, who have been drawn into an emotional relationship that—when the going is good—is deep, passionate, erotic, and rewarding do not know how to react when the magnetic stream of emotions is abruptly cut off.

The story that lies behind Moon-Pluto aspects is in a sense one of alienation that goes back a long way. Even the womb can seem like an alien environment, and later these people seem to have considerable ambivalence about the breast, which, on the one hand, represents succor and, on the other, something over which the child, much to his or her alarm, has no control. The life-giving milk can be withdrawn at any time. Power battles arise, perhaps because the mother feels that there should be some discipline or regularity but more likely because of the child's intense insistence on filling the security void in the stomach. For the generation with Pluto in Virgo, health and diet issues are often crucial. I have seen clients whose mothers thought enemas were the way toward perfect health, for example, which then sometimes gave rise to humiliation and odd anal fixations.

Sometimes the mental or emotional state of the mother is an issue for Moon-Pluto clients. Moon-Pluto clients are often born when the mother has been in a period of long-term crisis, or they are the unwitting contributors to her postnatal depression. They may have witnessed extreme emotional states in the mother, which no child can be expected to encompass, and are often victims of extreme behavior that issues forth from temporary madness in the family. Whatever the originating circumstances, people with this aspect feel the need, at some point, to find a strategy to handle what they perceive as a life-threatening emotional situation, and they do this by cutting themselves off from the world of emotions. The mother who puts her child "on ice" will find that the child learns the technology of amputation even better, resolving never to be vulnerable again.

However, the crucial issue here is that whatever the behavior of the mother, there were good reasons for it in the mother's life. These circumstances did not arise because the mother did not love the child—quite the opposite—the mother did love the child, but her situation prevented her from expressing it. Whatever behavior there was in the family environment, this triggered the child's latent survival instincts, and these instincts were to cauterize the area that hurt—in this case, the emotions. Later in life there is an instinctive drive to replay scenes of crisis and re-engage the ritual of emotional amputation. The client is in a position to do something about this now. Intervening in this vicious circle, the astrologer can point out this pattern, and it will be found that it stems from an indefinable anxiety centered somewhere in the body, most likely around the stomach and solar plexus area. It's the fear of void in this area that provokes a series of actions, which result in the same scenario of alienation over and over again.

Moon-Pluto is a grateful aspect. The very nature of Pluto is to bring unconscious issues up to the surface and transform them. Often some form of therapy, either individual or group, is the way forward. Certainly, it is a good idea for these people to develop the psychological skills that come so naturally.

What works for Moon-Pluto people is to be engaged in something that requires intense involvement and emotional commitment, transforming their environment or the lives of others. What is cathartic for them to admit is their unconscious need to manipulate others and the environment in order to feel safe and to be in control.

Mercury— Making Connections 6

If we close our eyes and attempt to still our mind for one minute, we will discover that this is a near-impossible act. Thoughts arise unbidden and the mind is never still. Even for the average meditation practitioner, the mind is only stilled for rare moments. Using a word like "mind" is very inadequate to describe what actually goes on with thought processes. Buddhism, for example, describes each aspect of mind with a specific expression—and there are many—just as Eskimos are said to have innumerable words for snow. Considering the limitations of language, then, Mercury can be said to relate to all forms of mental activity. The way you think, what you think about, the way you speak and what you say, the way you hear and what you imagine—these are all expressions of the aspect, sign, and house positions of Mercury in the horoscope. As soon as you open your mouth, Mercury slips out. The nuances of voice tone, tempo, intensity, and volume reveal with great accuracy what kind of Mercury this is. Any free association exercise will be a great vehicle for revelations about the nature of a person's Mercury.

When working with Mercury, it is crucial to take note of tone, tempo, and subject matter. Mercury-Mars in Sagittarius may be breathless and strident and talk angrily of justice; Mercury-Pluto in Virgo may be precise, intense, and clipped and talk anxiously about efficiency; Mercury-Venus in Cancer may be gushing and talk lovingly about home and family. Astrologers with a trained ear can identify how thought and emotion tinge communication, and they can act as a mirror to present clients with a new awareness of how their thinking is affected by their horoscope. Thoughts are subjective; we may be convinced of the truth of what we say, but the most we can ever say is that something is true for us. Our mental representation of reality tells us a lot about our mind, but little about so-called objective reality.

The influences on Mercury in horoscopes reveal much about the formative mental influences in childhood. These are primarily to do with siblings (or lack of them), their actions, and the way we relate to their actions. However, our school experiences also reflect Mercury, both in relation to academic achievement and as far as relationships with teachers go. More important still are our relations with our schoolmates and what goes on inside and outside the classroom in terms of friendships—or lack of them. This is often as much an 11th-house matter as a 3rd-house matter. Being teased or ridiculed, frozen out for shorter or longer periods, and being in specific cliques or groups—all these things have a powerful effect on our later development. Mercury-Saturn may be profoundly sensitive to ridicule, especially intellectually, Mercury-Neptune may feel isolated if not talked to, Mercury-Uranus may provoke boring teachers. We will all both evoke and be exposed to specific experiences according to our Mercury placements.

These early experiences will affect our communication and interests today as well as our attitude to truth and reality. When Mercury is seriously afflicted, it gives rise to painful mental disturbances, as this is the planet that connects us with reality. If the words we speak reflect a badly distorted representation of reality, then life will be filled with misunderstanding, and it will be difficult to achieve any long-term success. Mercury is very connected to how

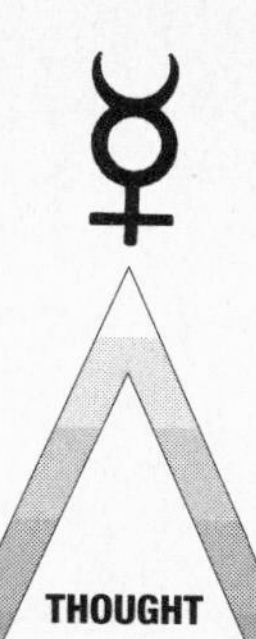

THOUGHT

Values:
Ever-arising thought processes. Knowledge. Ability to communicate. Genius, fantasy, skills.

Behavior:
Communication: inner and outer connections. Delivering the message, assembling information. Making contact.

Communication Polarities:
Truth/Lies. Quiet/Garrulous. Misleading/Direct. Secretive/Open. Critical/Supportive. Knowledgeable/Uncertain. Provocative/Diplomatic. Ignorant/Intelligent. Charming/Rude. Principled/Flexible.

In Society:
Postal services. Bicycles, cars, buses, trains. Communication infrastructure: telephones, e-mail, letters, etc. Schools, neighbors, siblings.

Body/Mind
Mobility, nervous system, hands and fingers. Mouth, ears. Nasal passages and lungs. State of mind. Mental balance. Inner dialogue.

Fig. 17. Manifestation of Mercury energy. Mercury's energy, combined with its sign position and any aspects, shows the style and content of communication, voice tone and delivery, and mental interests and attainments.

we get along in the world. Clearing up communication issues is a prerequisite for success, and, indeed, there are many excellent courses on the business market today that reflect this.

Fortunately, it is relatively easy for the astrologer to work with Mercury and communication issues, because these issues are immediately apparent in the words and attitude of the client. Whereas the Sun is related to existential matters and the Moon to unconscious ones, Mercury is the stuff of consciousness, and the issues connected with it can be rendered objective in the two-way process of communication. The inner dialogue is even more important than the outer; in the reverberating chambers of the brain, our inner conversations reign supreme and show how we *really* are. By guiding the client through these inner chambers, lasting change can be wrought.

As with all the aspect descriptions in this section, the following explanations are aimed at helping people for whom the aspect manifests in a strongly negative manner. A good number of people with stressful aspects may have already overcome many of the problems that the aspects suggest.

Mercury-Jupiter: Right and Wrong

There is no aspect like this one to give a hunger for knowledge and understanding. The process of learning can continue for the whole of life. It is not limited to the traditional years of education. Being intelligent is one of the highest criteria for people with this planet combination, and the kinds of problems that arise often occur when their intelligence is called into question. Generally, this is a gentle combination, and Jupiter has sufficient dignity to retain tolerance and diplomacy in discussions. With a bit of luck, the Mercury-Jupiter client has learned to balance the need to be right with the desire to acquire more information. However, if either is placed unfortunately, then there can be much argument in which Mercury-Jupiter uses considerable persuasive powers to put across its point of view. A strong Mercury in Scorpio would have considerable intellectual sting and probably harbor an old grudge, a strong Mercury in Aquarius would be impervious to argument, and so forth.

Afflictions connected with Mercury-Jupiter aspects are profoundly connected with a sense of injustice, particularly with regard to siblings or schooling. Often there is the sense that a sibling has had preferential treatment, and this can be the source of considerable indignation. The perception of injustice can easily arise even if the justification is small. To compound the matter, Mercury-Jupiter is inclined to be vociferous in pointing out any perceived unfairness, and is not content with compromise. In these cases, Jupiter desires nothing less than being right, especially if fire signs or the planet Mars are also involved.

During the consultation, if grudges are aired connected with friends, teachers, colleagues, or siblings, then this could be Mercury-Jupiter at work. The first difficulty the astrologer encounters is in creating any sort of perspective on the situation. The energy of this aspect manifests as a very persuasive inner voice that provides totally convincing arguments as to why Mercury-Jupiter is right. I recall a 70-year-old client with Mercury rising in Scorpio, opposing Jupiter in Taurus, who had nursed a grudge for many years because her sister received considerably more of the family inheritance than she did.

Examining the inheritance episode, two factors emerged: the sister was a lot poorer, and the client had previously expressed complete indifference to inheriting from her parents. She nevertheless used the episode as justification to add to the catalogue of sibling injustice that had been her lot in life. These clients *are* responsive to clever argumentation, though the arguments have to be marshaled in a masterly way, otherwise the astrologer will be on the losing end. Mercury-Jupiter, after all, has had a lot of training in being right.

The above example illustrates the point that it is the perception and attitude of the client that dictates how much she will suffer from the negative Mercury-Jupiter aspect. As the aspect gives a natural wisdom, it is often possible to nudge clients into a new perspective that liberates them from the tyrannical inner voice that tells them how much injustice they have had to put up with. They are particularly susceptible to a good "reframe," meaning that the act of putting the whole episode in another perspective can radically change their view. As this particular client prided herself on having carved out her own niche in life through her own efforts, what use would the large inheritance have had? Would she rather have seen her sister struggle through life without money? However much sibling resentment there may be, blood is thicker than water.

Square and opposition aspects are much more troublesome than the conjunction, which confers great intelligence and talent. The dilemma lies in the conflicting inner dialogue, with Jupiter expressing powerful convictions based on the sign and house it is in and Mercury constantly bringing up new arguments that conflict with these convictions. Mercury in Aquarius in the 11th might want to spend money on mental development courses; Jupiter in Taurus in the 2nd would have down-to-earth arguments as to why this is unwise, and so forth. In acute cases, Mercury-Jupiter people exhaust themselves in their inner dialogue, and when they finally plump for a course of action, new arguments immediately rear their heads, rendering them intellectually exhausted and confused. The crucial point is that thought is not action, thinking is not doing.

Exaggerated mental activity is just an expression of inner mental dynamics. In a world where meaning is a subjective concept, Mercury-Jupiter people exalt the idea of meaningfulness and fairness to unjustified heights, and dialectical considerations prevent them from making simple decisions. If you don't have the blueprint for Life (and who does?), events in the world are empty of meaning. Meaning is conferred on life by the minds of individuals, who are, in the last analysis, the mental architects of their reality. It's essential that Mercury-Jupiter clients are clear about what they make things "mean," and that they reorganize their convictions so that the meaning they derive from their interpretation of events is empowering rather than disempowering.

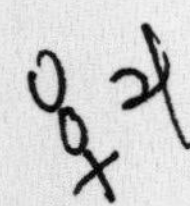

What works for Mercury-Jupiter clients is to be involved in activities in which they disseminate knowledge. What is cathartic for them to admit is that their inner dialogue is self-righteous and subjective.

Mercury-Saturn: Fear and Fluency

This aspect gives rise to a sense of insufficiency and lack of self-confidence connected with communication and intelligence matters. Saturn-Mercury people tend to be somewhat subdued and diffident, perhaps pedantic, and are careful not to expose any hint of ignorance. This can then manifest as defensiveness when they feel there is a chink in their mental armor. Saturn-Mercury says little about actual intelligence and much more about subjective fears regarding lack of it. The greatest hindrance to mental advance lies in this subjective fear and the energy wasted in concealing it. Saturn-Mercury is the portcullis that descends whenever there is a mental challenge that the person feels he or she cannot meet. In time, this can actually lead to an expectation to misunderstand and an elaborate cover-up to disguise the supposed lack of understanding.

The learning and communication difficulties associated with Mercury-Saturn basically come down to mental mechanics and can be solved by a re-engineering process. The client's awareness of communication difficulty probably started early in life. This could be connected to an older brother or sister who made a point of putting the client down intellectually, or indeed a younger sibling who was smarter than the client. Something in the family triggered the mental insecurity of Mercury-Saturn. The real difficulties began in school, however, and generally the client experiences a lot of secret shame because of perceived poor performance. Invariably, there was one specific teacher who made the client's life a misery. The client should have no difficulty supplying his or her name. There probably was a specific episode in which the client was exposed to ridicule in the class, and this is associated with a burning shame. At this key moment, the client made life-changing decisions on an inner level that dictated his or her process of intellectual development: a decision to work twice as hard—or to give up; a decision to avoid any exposed intellectual situation, to be invisible; a decision about mental inferiority.

During the consultation, you as the astrologer may become aware of some variation of Mercury-Saturn in the communication: a pretence at understanding without actually doing so; an excuse about being slow; self-deprecation; or a haughty, reserved, or dismissive facade that effectively wards off in-depth communication. This is the moment to intervene and to find out what is in fact going on in the client's internal dialogue. The mechanics of the process is usually:

- A sudden, inner gut feeling about being intellectually exposed, with accompanying self-critical inner voice;
- A mental block inhibiting intake of any further information that might enhance understanding;
- Initiation of a defense mechanism such as pretence, pedantry, or falsity.
- A deflection of threat but subsequent sense of being isolated from what might have been intellectually enriching.

All this could be avoided if Mercury-Saturn was to admit to difficulty in understanding. "Could you repeat that please?" or, "I'm sorry, I didn't get that," or "Excuse me, but I'm pretty slow. Could you spell it out please?" The trouble is that years of effort have gone into covering up a supposed inferiority. The client needs to know that it's not that serious to misunderstand—people don't care that much. But pretending to understand or cultivating mental barriers *is* serious. The Saturn-Mercury cure is far worse than the problem.

The astrologer must be supportive and understanding of the sensitivity the client feels in this area and slowly eke out an admission and description of the internal dialogue. This aspect responds well to training, and the way to begin is with creating a positive inner dialogue, rather than a negative one. Instead of Mercury-Saturn berating him- or herself for being dumb, the tone and direction of the inner voice must be guided into an appreciation of what the person can achieve.

The great thing about most Saturn aspects is that they improve with age, and especially after the first Saturn return at 29 to 30. The client resumes education that he or she abandoned in the teenage years or early 20s, and the Saturn that once created a mental block is now harnessed into a concerted long-term effort for mental training. Care must be taken to avoid using continued education as a crutch. At some point, the client should cast aside authority and develop spontaneity.

What works for Mercury-Saturn clients is to be involved in activities that require concentration and care, for they naturally tend to go over things more than once to be sure they are right. What is cathartic to admit is that their inner dialogue is self-dismissive and fearful and that they project a fragile facade of understanding, which they fear will be exposed as fraudulent.

Mercury-Uranus: Quite Contrary

When these planets work well together, the Mercury-Uranus person possesses remarkable genius in certain areas, especially anything to do with invention, experimentation, and the future. When the aspect is difficult, there is still considerable inventiveness, although the communication skills needed to enroll other people may not be well developed. Characteristic for Mercury-Uranus people is impatience at the slow pace of what is going on around them. When they communicate, they often try to liven up the people around them, even if this means being provocative. They may be impervious to the sensitivities of others, and though people can be fascinated by the novelty of an unusual form of communication, they can also be alienated or disoriented.

These types of client will generally be very open and curious about the astrological consultation—eager to learn about themselves via unconventional methods. Generally ahead of the game, they may have a tendency to interrupt and complete your sentence. Typically with Uranus aspects, there is no great sense of any problem to do with the aspect; nevertheless, behavioral patterns can arise that ultimately have negative consequences. One of these is the disruption of the communication process. In childhood, Mercury-Uranus often manifests itself as attention-seeking by being outspoken. In the competition for his or her parents' hearts and minds, the Mercury-Uranus child's ability to let loose an arrow from a very surprising direction becomes part of his or her arsenal.

Usually, during childhood, there was a key upheaval that put the child out of kilter. This could be something like changing classes, changing school, or the disappearance of a sibling. In extreme cases there is a severe event that totally blows out the cognitive processes. One participant in one of my courses had Mercury in Sagittarius in the 2nd house, opposite Uranus in the 8th, the latter conferring circumstances likely to be traumatic.

She could recall nothing of importance, despite considerable investigation of childhood memories. During the lunch break, she saw sparks flying from a tram that went by, which triggered the most extraordinary memory, utterly suppressed since she was 16. At that time, she was committed to a mental hospital and was given electric shock treatment, but mistakenly without the normal anesthetic to remove the pain—an enormous trauma. It was only at this course, age 40, that she recalled the event. This example, extreme as it is, illustrates the short-circuiting tendency to which Mercury-Uranus aspects can give rise. It is as if there are holes in the memory, especially if there is also a Plutonian influence.

One characteristic of this aspect is the client's tendency to go off on a tangent in conversation. He or she has a belief that there is more interest to be derived from the distraction, that it's a more exciting area than the actual subject being discussed. Communication stops and starts, and the other person has difficulty getting an argument across. Unpredictability may be a means of being in control of the conversation, and it may also be a manifestation of fear of "boredom." The conventional and the predictable are anathema to Mercury-Uranus, which is one of the reasons that these people manifest inventive talent. Yet "boredom" is a loaded word, a cover story for hidden fears. Mercury-Uranus clients tend to be "airheads." They live in a world of exciting ideas but are removed from the reality of the here and now. To be in the body, at one with the senses and in close contact with physical reality, can be very difficult for them, and the result of this is an indefinable feeling of remoteness, a sense of being out on a limb. This is a very real result of communication habits that are disconnective rather than connective.

As the consulting astrologer, you may be able to spot these communication patterns in the consultation if they are pronounced. The signs to watch for are the client's restlessness, impatience, and tendency to interrupt you. When Mercury-Uranus clients interrupt, this can be a signal that they are trying to steer the conversation away from an uncomfortable area, and it will be disempowering for you. It's a perfect opportunity for an intervention. "What exactly was I going to say, then?" If a client interrupts the astrologer, it's a sure sign that he or she will interrupt others a lot, and this will create disharmony in relationships.

What works for Mercury-Uranus clients is to be at the forefront in a constantly changing and stimulating mental environment, preferably with an international or new-age dimension. What is cathartic for them to admit is that they have little control over a provocative inner dialogue that distracts them and prevents them from tuning in to others.

Mercury-Neptune: The Smokescreen

Attunement to subliminal stimuli is one of the strongest talents of Mercury-Neptune, whether the aspect is harmonious or disharmonious. The person with this aspect has tremendous imaginative power and sensitivity in communication. At an early age, the boundary between the real and the imagined is blurred, giving rise to confusion about what is truth and what is not. In communication, this person has a strong awareness of what is going on in the mind of another person and a tendency to adjust communication correspondingly. Mercury-Neptune listens acutely to what is said, and what is unsaid, and has difficulty distinguishing between its own thoughts and the thoughts of others. The Mercury-Neptune person has a chameleonlike quality of taking on the surrounding mental atmosphere.

In the consultation, Mercury-Neptune clients absorb what is said by using a process of free association, allowing words to wash through the mind and trigger ideas and memories. They are "mind readers," divining the meaning of what is being said before it is complete, often coming to an understanding that least disturbs their preconceptions. Therefore, it is crucial to check that what you say and what the client thinks you have said are the same thing. They often aren't. This habit—combined with an unwillingness to say anything that may create bade vibes—naturally leads to many misunderstandings in life. There is a discrepancy between what they actually say and what they are in fact thinking. This discrepancy can often mean that their words lack real power, and they can often feel as if they are communicating through cotton wool or that other people do not pay enough attention to what they say. They can find it very difficult to express themselves when under pressure and have great difficulty formulating words that adequately express what is going on inside them.

In early childhood, the client may have created a secret reality of untruth. The child may have learned early that it is better to be evasive than frank. He or she may prefer to lie to avoid unpleasantness or in order to have a bearable life, and in the long run this creates an environment that the child navigates very skillfully but that necessitates maintaining a false world in which he or she has the capacity to completely believe. It is extraordinary how, later on in life, Mercury-Neptune people can actually believe their version of reality, despite overwhelming evidence to the contrary. In their inner dialogue, they avoid the unpalatable and believe what they want to believe. This can sometimes manifest as real dishonesty, not because of any malice but rather because they have convinced themselves that the fabric of reality they have created is objectively true. Another aspect of the Mercury-Neptune childhood is a sense of isolation, both in the family and at school, or suffering from a situation of being overlooked or ignored. This stimulates the need to retreat into an inner world of the imagination.

One of the signs of this aspect in the consultation is when you, as the astrologer, lose track of what is happening or get confused. Mercury-Neptune clients instinctively create a smokescreen behind which they can retreat to prepare a new convincing story. This leaves you disoriented, as indeed anyone in Mercury-Neptune's intimate circle will inevitably be. It is important to use language very precisely, as described earlier in chapter 3, so that you can pin these clients down and encourage them to elucidate. As they get deeper into the meaning of what they are saying and thinking, they become aware of an indefinable sadness in their being. Perhaps it is the sadness of being misunderstood, or unheard, or perhaps the sadness of separation between humans. Neptune seeks to dissolve the boundaries both within the mind and between people. Yet, to function well in the world, mental boundaries are essential. Mercury-Neptune needs to encompass that feeling of inner loneliness and from there develop attunement to others.

What works for Mercury-Neptune clients is to be engaged in areas that require an intuitive sensitivity, where they are undisturbed by noise, hustle, and bustle. What is cathartic for them to admit is that they feel alone and ignored, that they cover up the truth, and that they fear their ideas carry no weight in the world.

Mercury-Pluto: Secrets and Lies

People with this aspect are born detectives. They have the ability to divine what is going on beneath the surface and the desire to probe others' secrets. With Mercury-Pluto, they have a special mental filter in operation that makes them question what is said, because they suspect that something more important is left unsaid. During the consultation, Mercury-Pluto people often feel that you are concealing some dark truth from them and urge you to tell all. Delving as they do into the secret motives of others, they will often find that important facts and events have been concealed from them, confirming their conviction that you can't trust people too far. People intimately involved with them often feel under interrogation and resent having their motives questioned. Paradoxically, this can lead to them concealing information, especially if they feel that the Mercury-Pluto person will react adversely to hearing the truth.

Mercury-Pluto clients themselves are adept at concealing their own thoughts and even the unconscious, subliminal signals that generally reveal inner thoughts, such as eye movements and body language. They have a tendency to transfix you with eye contact, gaining psychic control. Their voice tone can be monotonous and have a hypnotic effect. Many people with this aspect have a way of droning on and completely exhausting the listener. When this happens, the listener feels the urge to break off communication, but doesn't dare. Under extreme circumstances, other people will avoid communication with the Mercury-Pluto person precisely because they feel dominated and exhausted and because there is no two-way flow in the communication.

What is really happening is that the Mercury-Pluto person feels anxiety when communicating. They transfix the listener for fear of communication being broken, only to find they are avoided in conversation, which is what they least want to happen. They keep talking because they fear the silence when they stop. This is the crucial point at which the astrological intervention can take place. If Mercury-Pluto clients go on and on, then you, as the astrologer, should point this out, even though this may provoke a strong reaction. The secret of transformation here is in the empty abyss of silence when the client ceases to fill the space with words. These clients must learn to deliver their sentence and then pause and wait for a response. In the pause lies a pool of anxiety, which *is* the Mercury-Pluto energy. To linger in this space gives the client access to a transformative power, even though it is uncomfortable at first.

One of the characteristics of the Mercury-Pluto childhood is the charged atmosphere of what is unspoken. There are almost always key secrets in childhood—secrets that can poison perception years into the future. Sometimes these secrets are not exposed until quite late in life. Often there are half-brothers or sisters, and sometimes the child will be unaware of this, or—if aware—will not be allowed to speak of them in certain environments. Yet, with the gifts of detection that Mercury-Pluto brings, the child senses that something is concealed, some areas are taboo, and some things must not be talked about. It is extraordinary that what is concealed has so much more power than what is spoken of. Apart from this atmosphere of secrecy, the child is often subjected to cruel practices in communication, notably when a family member deliberately shuts the child out for long periods and refuses to utter a word, often as a form of punishment.

Another characteristic is the tendency to brood, or, more precisely, to engage in exhausting inner discussions in which minor issues are analyzed in depth without coming to any conclusion. When in an intimate relationship, this inner discussion is often projected, so that the partner is drawn into interrogation-style conversations that go on long into the night. Mercury-Pluto hopes to find "the answer," and does not know when to stop. There is no answer. The partner has to find a strategy for survival under this unrelenting stream of language. Two things can happen: either the partner gets involved in long arguments or shuts up completely. The astrologer will almost invariably find these clients complaining that at times their loved one hardly talks to them or is quiet for long periods. The Mercury-Pluto client must resolve to deliver the message, and then shut up. No good comes from pursuing the matter, only harm. It all boils down to the inner feeling that some truth will emerge that will bring release, but—as this feeling stems back to the beginning of memory—the idea of release is an illusion.

What works for Mercury-Pluto clients is to be engaged in areas that require analytical and detective abilities, especially connected with psychology, computers, and anything that requires delving beneath the surface. What is cathartic for them to admit is that they greet much of what is said with suspicion. They expect people to conceal things from them, but they actually conceal things from others.

Venus— Bonding 7

If the Sun represents the blinding light of consciousness, the Moon the comforting cradle of the unconscious, and Mercury the ceaseless play of thought creating connections from inner to outer, then Venus is the stage of existence where shape and form are created in an expression of harmony reflecting the life force. Whenever we are receptive to a sensory stimulus, we instinctively form judgments as to whether we like something, dislike it, or are indifferent to it. This happens rapidly and unconsciously—we attach labels to people and things and treat them according to our likes and dislikes. Venus is the yardstick by which we measure what attracts us and what doesn't. It is also the yardstick by which we judge ourselves to be of value or not. But, in fact, people and things are what they are; it is we who make judgments apportioning value to things. This value is not inherent. It is relative to the experiencer and his or her situation. Water, for example, is much more valuable than gold if you are alone in the middle of a vast desert.

The sign and house position of Venus, along with its aspects, show precisely how our value judgments function. As life unfolds around us, we constantly apply our Venus yardstick to see how others and we are measuring up. Venus in Virgo will see the dust on the floor and the accumulated dishes and judge accordingly, Venus in Aries will not. Venus-Saturn combinations will measure a relationship from the yardstick of security, Venus-Uranus from the yardstick of excitement. This constant evaluation of the surroundings creates a kind of magnetic field or aura that filters events, objects, and people, actually preventing us from seeing them for what they are. But it's a natural process, and we gravitate to certain environments and social circles according to our Venus filter, choosing precisely *that* picture to go on to the wall, *this* pair of shoes, *this* partner, and so on.

The influences of Venus in our horoscope tell of how we experienced love as a child, and how this affects our self-worth and evaluation of others. Venus particularly relates to sisters. A description of Venus will often describe how the client experienced his or her sisters or, in the event of having no sister, how girlfriends were. Venus will also tell of the love or lack of love as a child and particularly how the parents expressed love. Venus shows how we let love in: if it is configured with Jupiter, joyfully; with Saturn, carefully; with Uranus sporadically; with Neptune, in our dreams; and with Pluto, with life in the balance.

Early experiences of love naturally affect how we give and receive love today, and the values and habits formed as a child will inform our choices today. A seriously challenged Venus will lead to specific experiences in love that are almost entirely due to the values and decisions made in the first few years of life. When the fear of being abandoned or a craving for appreciation influences our behavior in relationships, the effect on our partners will be counterproductive. In a society that has lost the taboos around marriage and separation, it is amazing that some couples do stay together, given the difficulties inherent in so may Venus placements. To have a rewarding relationship today, it is essential to understand the role Venus plays in the chart and to consciously work to optimize its positive sides and minimize its negative sides.

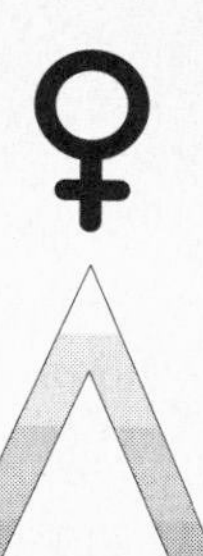

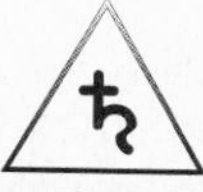

Fig. 18. Manifestation of Venus energy. Venus's energy, combined with its sign position and any aspects, shows basic tastes, leading to attraction to specific people and things.

There is a profound difference between Venus in a man's horoscope, and Venus in a woman's. My one-year-older sister was careful to inform me that boys were made of "slugs and snails and puppy dogs' tails," while, as a girl, she was "sugar and spice and all things nice." The gender conventions of the society children grow up in are very dominating. The values of little boys and little girls are, or become, radically different. My sister made little effort to repair *my* bicycle, and I just could not get into dressing her dolls. Boys identify with Mars, and girls with Venus—that's the cultural picture today, though there is more crossover than in previous centuries. As such, aspects to Venus in a woman's chart go straight to the heart of her identity as a woman, and they are crucial to whether or not she feels attractive.

In a man's chart, Venus does show something about his self-worth, but worrying about this is not consciously a man thing, and this self-worth tends to be projected. Therefore, description of Venus in a man's horoscope will seem to reflect the kind of woman he is attracted to; Venus in Leo will relate to wanting a female partner who creates a good impression, for example. This does not mean that the woman is really as the man experiences, it shows rather what behavior he evokes in her—what he lets through the filter and what he shuts out. In all relationships, we tend to show behavior that the partner approves of, and as a result, we can be one person with one partner and quite another with the next.

Working with Venus in the consultation is slightly easier than working with the Sun, Moon, or Mercury, because its effects so obviously manifest in relationships, and clients can easily identify the repeated love experiences described by the astrologer. Issues of self-worth, though very sensitive, lie just under the surface, and if the astrologer establishes a good rapport with the client, then he or she will be willing to bring any fears to the surface. Once the client recognizes counterproductive behavior in relationships, it is not difficult to enroll him or her in new patterns of behavior that will bring positive results. Venus-Pluto can easily learn that the love autopsy is a bad strategy in a living relationship; Venus-Neptune can buy the idea that they must love the one they're with, and so on. With Venus, the astrologer is working on evoking *feelings*. Whereas Mercury relates to inner dialogue, Venus relates to inner feeling—often a quite physical sensation in the body, and often in the throat area. This is where sensations connected with love and hurt are stored, and this is where they need to be accessed to be dealt with in the consultation.

Venus-Jupiter: Horn of Plenty

This combination can be one of the most fortunate in terms of pleasure and happiness, particularly with the sextile, trine, and conjunction, and if both planets are placed well. From its Taurus side, Venus gives the potential for sensory pleasures and satisfaction, and from its Libra side, the joys of social interaction and aesthetic or cultural stimulation. Even the square and opposition will not necessarily result in painful experiences, unless either are placed badly. Venus is most attracted to wisdom with this placement, so there is a strong need to find a partner who can be admired for being cultured and influential. Venus-Jupiter is drawn to people from foreign countries—often neighboring countries—and to people with enlightened philosophical views. The opposition aspect, more than any other, indicates relationships conducted from afar, with marked cultural differences.

One area of difficulty with the conjunction, square, and opposition is *surfeit*. There is a considerable appetite for love and pleasure, which can lead to serial relationships when the appetite for love is satiated by one partner, and the hunger for more love sets in. This can also be reflected in eating and drinking habits. When Jupiter afflicts Venus, moral issues also come to the fore. It is no coincidence that Vedic astrologers portray Venus as a "demon," along with Mars, Saturn, and the Nodes, while Jupiter is in the pantheon of the gods, along with the Sun and Moon. Venus is not "spiritual" in itself, and the love associated with Venus is not necessarily a spiritual love but rather the satisfaction of needs connected with pleasure and satisfaction.

It's difficult to bring moral considerations up to the surface in a consultation. You can only point to consequences and see whether clients can identify them and get enrolled in changing behavior. If Venus-Jupiter does manifest as serial relationships and clients wish to change this, then they have to address their inner need for romantic stimulation and look at their urge to feel that their partner has something to teach them. Early in a relationship, Venus-Jupiter people will respect the partner for the cultural breadth they feel they themselves do not possess. As time goes by, cultural and aesthetic values are absorbed from the partner, and then perhaps there is a feeling that there is nothing more to learn. At the same time, the heady romantic days with meals at fine restaurants, exotic holidays, and an expanded social circle will ebb as the relationship settles. It is at this time that they can be seduced by the forbidden fruits of new love. The partner can only satisfy intellectual needs for a limited time, and ideally the Venus-Jupiter client should be the one to pursue cultural and intellectual development through his or her own initiative, rather than projecting the need onto the partner.

Venus-Jupiter has the highest expectations of what a relationship can provide. These people give effortlessly when in love and, consequently, also evoke generous and romantic behavior from their partners. For them, ugly behavior is not an option, and when difficult times arrive and there is anger, cruelty, or argument, they feel that the central core of their values has been sullied and are indignant that their generous behavior should be so unfairly rewarded. But if there is a refusal to face up to unpleasant truths—like extravagance leads to poverty, indulgence to surfeit, promiscuity to loss—then, in the end, the partner can be forced to spoil the party, and the romantic dream leads to a rude awakening. The partner may not be forgiven for being so small-minded. Venus-Jupiter people find it almost impossible to live with ugliness or ignorance in the surroundings. A disagreement on matters of taste, which could be a minor issue in other types of relationship, can threaten the very existence of the Venus-Jupiter relationship. These people must have their own way in taste, and they have no time for philistines.

If these clients are having relationship or economic troubles, then it will at least partly be due to the behavior described above.

It would be unrealistic for these clients to expect constant satisfaction in love, and if the partner is seen as boorish or demonstratively unromantic, it could be because this person tries to be realistic in the relationship and economically responsible. In fact, in doing so, the partner may be providing the necessary balance in the relationship, and it is to the advantage of the client that this be so. If the partner did not fulfill this role so disdained by Venus-Jupiter, who would? Will another partner not find the Venus-Jupiter client evoking similar behavior?

Notwithstanding the possible difficulties I just described, Venus-Jupiter clients confer many wonderful blessings on those they love. What works for them is a stimulating, intellectual relationship with someone they can look up to for guidance. What is cathartic for them to admit is that pleasure has a price. Happiness in love is not an inherent right; life is not a Barbara Cartland novel, and disappointments often stem from their own spoiled behavior.

Venus-Saturn: Hungry for Love

Contacts between Venus and Saturn bring an atmosphere of formality to relationships, often reflected in a polite and attentive manner. Venus-Saturn people send the message that they are prepared to live up to their relationship commitments. In social gatherings they create a good impression, but when all the guests have left and they are alone with their partner, they feel more clumsy and uncomfortable. A lot of the difficulties these people encounter are a consequence of the endeavors they make to avoid getting hurt in love.

It's a fairly safe bet that a key episode took place at around the age of 15, when the Venus-Saturn person felt rejected by a first love. This is partly because it is around this age that the person sends out romantic feelers and embarks upon intimate relationships and partly because Saturn makes an opposition to its natal position at this time, triggering the Venus-Saturn aspect. Where other young people may take a rebuff with equanimity at this time, Venus-Saturn people draw conclusions and make decisions that can affect the rest of their life.

They are slow to reciprocate affection, and potential partners may conclude they are not interested and search elsewhere. Sometimes Venus-Saturn people choose a relationship for safety rather than love, opting for someone who loves them but for whom they merely feel affection—something that can have consequences later.

There would seem to be a karmic element involved in some of these relationships, which may involve shouldering some heavy responsibility. In a sense, the karma is the cure. One female client with Venus rising opposing Saturn had a child who had a rare copper (Venus!) deficiency, which meant she had to nurse her daughter constantly to maintain her in good health. This evoked great love and concern, which was a purifying factor in the client's life.

A man with a hard Venus-Saturn aspect will often complain that his partner is an unhappy and depressed person, and his relationship can be very difficult, indeed. However, sometimes the sorrow of the wife is due to the lack of love she feels, either because the relationship is more of a safety measure than a love match or because the man is unable to express love in a warm and spontaneous manner. Such a man has an enormous reluctance to "put out," to make a declaration of love without a safety parachute. If the male client expresses discontent with his partner, it is crucial that he understand that he is in a position to influence her by transforming stinginess into generosity, even if this means opening up to the possibility of rejection.

A woman with this aspect is convinced from experience that she is not attractive, even if she really does look good. Even when she hears and feels declarations of love, it is not enough to convince her of her desirability. The problem lies with the inner feelings, which no amount of outer experience can change. These feelings have their root in the perception of being unloved from a very early age. There can be many reasons for this: the parents' marriage may have been difficult; the mother may have projected expectations on the child that the child could not fulfill; she may have had painful experiences at school, difficult experiences with a sister, and so on. The point is, however, that the Venus-Saturn filter ensures that the interpretation of events concerned with love is rather pessimistic. Venus-Saturn people feel the coolness between the parents but may miss the warmth.

It is worth challenging their interpretation of their childhood in an effort to eradicate convictions about self-worth that may be based on a wrong assessment of the situation. For example, the client may say that his or her mother never showed affection, she was too busy. Let's say the Venus-Saturn client described her as a hard-working, professional, single mother. What a job! Bringing in the money *and* caring for the child. That's also a way of showing love. And—as Venus-Saturn really shows a client who values professionalism—what a great example for the child! Reframing events so that the client sees them in a completely different light is a very effective way of getting rid of the disempowering convictions that the client may have. Appreciating the efforts that other people have made, despite the difficulties, is a big step toward self-appreciation, which is the antidote to the lack of self-love a client with this aspect can experience.

The painful lesson of Venus-Saturn is that if love is demanded, it will be withheld. Self-sufficiency has to be developed founded on an inner conviction of one's own worth. From that foundation, love can be given without fear of rejection. What works for Venus-Saturn clients is to have a rewarding professional life and a stable committed relationship. What is cathartic for them to admit is that, due to fear of getting hurt, they are stingy with love.

Venus-Uranus: Birds in Flight

Venus-Uranus aspects give a kind of restless sociability and great curiosity about other people, especially those from unusual and exotic cultural backgrounds. As Venus is the force that relates to the aura of attraction around a person, the combination with Uranus generates an extraordinary electricity and magnetism that can be quite bewitching. This in itself can be disturbing in a traditional relationship with values based on domesticity and exclusivity.

Usually, Venus-Uranus people fulfill their need for excitement and interaction by having a very active social life alongside a traditional relationship. If they were forced to choose, they would not want to give up their friends.

In the consultation, Venus-Uranus clients tend to create a good impression—lively and alert, very interested in astrology, and open for change. They may tell you about a number of relationships, none of which settled down into a long-term commitment, and although they may miss intimacy, it will not seem to be the highest priority. Their interests may well have taken them to foreign lands, and it's an odds-on bet that they have had a relationship with someone of another nationality and perhaps from a completely different culture. In these cases what is happening is that the need for excitement and originality, and for every day to hold something new, is satisfied by a partner who is so foreign that it takes years to experience the relationship as normal. The danger signs appear when a relationship settles into a routine and the partner becomes predictable. When this happens, the Venus-Uranus person becomes restless and has the tendency to provoke—anything to make something *happen*.

It is often the case with stressful aspects between Venus and Uranus that a friend was lost early in life. Sometimes it is a sibling, most likely a sister, who went off to boarding school, for example, or to stay with the separated father. Surprisingly often, the mother has had an abortion or lost a child a year or so before or after birth. Sometimes it's a good friend who moved on or parents that separated. Men with this aspect are often brought up in a rather androgynous way, and there can be some confusion about gender identity, especially if the Sun is involved with Venus. In the latter case, the son may have been treated as a girl, perhaps because of the prior loss of a child.

For Venus-Uranus clients, childhood separations are a sudden reminder that bonds can be broken and that intimate relationships can be ripped apart. To their eyes and ears it seems that these ruptures in human relationships are treated with matter-of-factness and reasonableness that make the extreme appear almost normal.

The faculties of reason place emotion at a safe distance, and a pattern of relating arises that tacitly accepts that a relationship can end, and hey, that's OK.

For people intimately involved with them, this can be infuriating. The message from Venus-Uranus is that if emotions are going to get all churned up, then maybe it's not worth continuing, and anyway, if it's going to be that much trouble, then let's cool down . . . we can always be friends. It's all sweet reason, and the Venus-Uranus person has a sense of being more in control—more enlightened—than the partner, who by now, quite unreasonably, is frothing at the mouth. But the big question is what is Venus-Uranus *feeling?* The emotions may have been cauterized, and this may result in an absence of feeling, an ice-cold sensation, which, even for the Venus-Uranus person, is eerie. There is an all-too-familiar feeling of zooming out into the stratosphere and experiencing everything from a great distance. The feeling here is isolation or alienation.

So while Venus-Uranus clients may feel it's the partner who creates the problems by getting emotional, they *will* be able to identify the feeling of distance and clinical coolness that is their emotional reaction. This is the entry point for the astrologer in the process of transformation. Once in contact with the feeling, it can be isolated and strengthened by techniques described in part three. This feeling is timeless, and being in contact with it means that clients can easily access early events where a mental break was made in a deeply emotional situation. The point is that Venus-Uranus clients do have very strong emotions, just like anyone else, but they avoid hurt through a lightning-fast process of disassociation. If they are unable to get into contact with their emotions, then they can expect the predictable pattern of Venus-Uranus relationships: sudden attraction, intense excitement, boredom and restlessness, then finally separation.

What works for Venus-Uranus is to be involved with work in which they have many abrupt and exciting contacts with people during the day, preferably in an unusual environment with international connections. They may prosper better in a relationship with someone who does not come from the same milieu as they do, and a relationship characterized by travel or short separations would suit them. What is cathartic for them to admit is that behind a friendly and reasonable facade they are not emotionally in touch with their partner or themselves.

Venus-Neptune: Longing for Pure Love

The combination of Venus and Neptune brings a spiritual quality and deep inner longing for union. Venus-Neptune people tend to exude a loving atmosphere and a wistful aura. They are attracted to idealistic communities and they seek to express the ineffable through art and other creative ventures. They have a talent for capturing universal aspects of love and romance in their work, and they can also awaken these archetypes in others. As Venus has to do with love and attraction, and Neptune harnesses discontent as a process of spiritual refinement, it is very difficult for these people to accept relationships for what they are. They always dream that the perfect relationship will come along. In fact, if the astrologer suggests to them that the perfect relationship will never come along (because of the peculiar dynamics of Venus/Neptune) the client will greet the remark with utter disbelief. You can't take that dream from them.

Venus-Neptune clients are especially attracted to creative and musical people, who manifest in their work the universal qualities of love through creativity. However, the typical pattern in a relationship is that—after the glorious Utopia of falling in love and merging of the soul—disappointment and disillusion arise. There is an extraordinary perfectionism with Neptune aspects, and in this case, it seems that after a few months elapse, Venus-Neptune begins to focus on all those things that *aren't* perfect with the loved one, and human nature being what it is, there is a lot of grist for that mill.

If you check carefully, you will discover that some Venus-Neptune clients nurture an ideal relationship in a secret fantasy world. They are very reluctant to reveal this precious dream, but it is essential to get to it. Even if involved in a long-term relationship, they may either have a dream of some long-lost love or some potentially perfect future relationship. Having great imaginative power, they create a very clear (but totally unrealistic) picture of this person, and spend considerable time living in this dream. It becomes the yardstick for measuring the failings of their partner.

Naturally, this does not elude the partner's awareness, if only on a subliminal level. Men with this aspect will often complain that their partner is sad or depressed, but they fail to make the connection with their own behavior. They unknowingly send the message that they are dissatisfied with their partner, and this naturally undermines her confidence and self-worth. Even a confident and self-assured woman can be filled with self-doubt after a year or so, especially as the relationship started with wonderfully affirming declarations of undying, spiritual love. If it turns out in the consultation that the client does indulge in fantasies about another person, then the consequences should be spelled out. While not intending to, they are completely undermining their existing partner. And, if that ideal person did turn up, would they then be satisfied? Almost invariably, Venus-Neptune clients will manifest in their own behavior the faults that so disappoint them in their partner, and this can normally be established through incisive communication techniques.

The wistful longing of Venus-Neptune has its roots in early childhood experiences. With particularly heavy placements of Venus, there may have been a genuine tragedy or loss in childhood that profoundly affected the client's view of love and attachment. More often, however, the aspect is connected to the client's perception of the parental relationship. One of the parents, usually the mother, may have given the impression of being dissatisfied with the other. Often the child is initiated into a maternal secret—a love she had once had but lost. The message is that the relationship you have got is not as good as the relationship you could have had. Venus-Neptune tends to filter out the signs of contentment in the family and focus on the unfulfilled romantic fantasies, tuning in to an aura of sadness. This sensory state becomes very familiar and is cultivated because it holds the promise of perfection. That is why there is such a strong reaction when the astrologer points out that there never will be a perfect relationship—it's like taking away a part of the personality. But how can true happiness ever arise unless the client sees through the veil of illusion?

When Venus aspects the outer planets, there is always a factor in love that means it can only be satisfied above and beyond the confines of a one-to-one relationship. Because this aspect gives a strong awareness of universal unhappiness—which is what the client is actually tuning in to and mistaking for personal—there is an identification with, and compassion for, people or groups with whom there is no particular connection. These feelings often lead Venus-Neptune individuals into spiritual, musical, or artistic areas that can be immensely satisfying. If they can be fulfilled in these areas, then the strain is taken off the personal relationship.

So what works for Venus-Neptune is to have an area of creative interest that they can develop alone and to have a partner who is also creative and satisfies as much as possible the highest romantic criteria. What is cathartic for them to admit is that while professing the very highest ideals, they are inauthentic in their relationship.

Venus-Pluto: Love and Survival

These aspects give intense magnetism and an erotic aura, which exert a deep fascination on others. Venus-Pluto people are concerned about the impression they make and can sometimes invest enormous energy into being liked. Generally, they successfully enroll people into liking them through a kind of relentless charm, and as the astrologer, you will be on the receiving end of this. In the short term, it can be very pleasurable, but long-term partners will learn to identify the anxiety that lies behind the efforts to be loved. Jealousy can plague their early relationships, as they feel they are at the mercy of another's love and cannot control whether it will be given or withheld. These people give intensely of themselves, and there is no way their relationships can be superficial, but at the same time, they draw the emotional energy from others and leave them exhausted.

The general pattern for Venus-Pluto clients is that they get involved in emotionally demanding relationships with a strong sexual element. They are very anxious about being rejected and therefore will stick with the partner through extreme situations of great psychological intensity. Their long-term relationships can be quite satisfying for long periods, then slowly rise in intensity, reaching crisis proportions in which everything appears to be lost. But Venus-Pluto people ride the storm, profess undying love, make up to each other in erotic union, and get on with the relationship. In this process there is a tendency to reveal the inner psychological state with ruthless honesty—the facade of charm is ripped away, and their "true" nature is shown just like a Strindberg drama. The problem is that this psychological striptease slowly undermines the relationship; each crisis is another nail in the coffin of love. Revealing this shadow side turns out to be very unwise; the client should avoid doing this, because no partnership can bear it in the long run.

Issues of power are paramount with these aspects—particularly with the Venus-Pluto square. Often men will try to attain safety by exerting control through the ability to turn love off and on, depending upon how the partner behaves. If the male client remarks that partners or ex-partners consistently have exhibited paranoid tendencies and psychological instability, it is probably a direct result of this power tactic. Being in a relationship that swings from extremes of passion to a barren, ice-cold landscape is enough to reduce anyone to a husk of what they once were. Women with this aspect will often get into one-way relationships where either she does not love someone who is desperately in love with her or really loves someone who is not interested. If the latter is the case, then the woman is open to complete humiliation, and though she may be interesting as a sexual prize, she runs the constant risk of being dumped. To avoid this eventuality she may develop behavior that actually invites her partner to treat her as a door mat. Her self-worth can be so low that contempt is awakened in the partner. Venus-Pluto people have to particularly avoid being compromised economically in relationships, as this later becomes the battlefield where power games take place. Both sexes are drawn to the "forbidden," yet a relationship casually embarked upon releases the destructive power of Pluto, leading to complications nobody bargained on.

The energy core of Venus-Pluto is the relationship between love and survival. Not being loved is unconsciously considered life-threatening, and this goes far back beyond memory, often as far as the womb when the mother entertained ideas of getting rid of the child for one reason or another. It is from this deep source that the urgent need for love stems. Other extreme situations are quite likely in childhood, including the cruel use of affection as a manipulation tool. It does not escape the attention of the family that the child is desperate for love, and from there it can be a natural step to give and withdraw love to insure compliance. As protection, the child learns effective emotional amputation techniques to beat those he or she loves at their own game. That is why in later relationships the Venus-Pluto person can suddenly change from being very affectionate to being void of love, feeling nothing. The partner can detect no love at all and wonders who he or she has been involved with all this time. But the Venus-Pluto client feels dead inside and is in pain.

This energy of deadness is the sensory state that can be effectively harnessed in the astrological consultation. Carried on this energy wave, Venus-Pluto clients identify a string of events that caused them to amputate from others and learn there is a reason behind the behavior. There is an in-built transformative power here, which can lead to empowerment and a new way of relating. These people have to learn to turn down the intensity and turn down the charm. Experience shows them they will *not* die if they are not loved, and if they relax, then they will be valued for their depth rather than feared for their jealousy, power, or control.

What works for Venus-Pluto is to be involved with transformation or psychological issues and to have a deep relationship with a committed and faithful partner. What is cathartic for them to admit is that they are secretly paranoid about their partner leaving them and that their love and sexuality are measured to keep control of the relationship.

Mars—Warrior and Lover 8

While the inner planets and the Sun and Moon represent states of inner being that closely interact and form the basic energy of Self, Mars—as the first planet beyond the Earth—is a driving force outward that powerfully interacts with the environment. It relates to the basic survival instinct, fight or flight, win or die. Where Mars is in the chart, there is always trouble. Mars is pure dynamic energy, and though it can be channeled, at its root it is primitive force. It can be constructive, and it can be very destructive. It ignites the flame of passion and desire and acts out the doctrine of the survival of the fittest. If Mars is too weak, then the person will back away from conflict, give up when meeting resistance, and let aggression out in perverted ways. If too strong, the person will bulldoze others without concern and make every meeting into a confrontation. However, the energy of Mars is the essential ingredient of achievement, whether material or spiritual. Harnessing Mars effectively is the secret of success.

For a man, Mars goes straight to the heart of his identity. The defining moments are the fights he has with other boys at school—both on and off the playing field, whether he wins or loses—and how he reacts when he wins or loses those fights. Mars is the experience he has of male figures, his father, and particularly his brothers. Mars's sign and house positions reflect how these figures measured up and what effect this had on the child's masculine self-image. Was the father henpecked and harried by the mother, or did he rule autocratically? Did he show his emotions, did he show rage, did he show tears? Was there an older or younger brother; was he kind and supportive, or cruel or jealous?

When it comes to the expression of sexual energy, Mars is the key player, both for men and women. If you are going to work on difficult Mars aspects in the consultation, you are going to have to talk about sex, and often in quite some detail. Otherwise, you may as well not bring the subject of Mars up. Even as a male astrologer working with 70 percent female clients, I have found it easier than I first imagined. (In fact, men are far more reluctant to talk about sex than women are.) Behavior associated with Mars will have specific consequences sexually, and when these are identified, curiosity displaces modesty and the client will want to know how things can be improved. If a relationship isn't functioning sexually, at least until middle age, then it is in danger. It would therefore be irresponsible not to offer help and advice. If astrologer and client are of different sex, then there is a unique opportunity to give advice that the client probably could never get otherwise. An astrologer adhering to clear ethical principles about noninvolvement with clients will convey this easily, and the client will know he or she can be trusted. If the astrologer or the client feels uncomfortable with this subject, then there are methods to deal with it without delving into private detail by working with inner sensory representations of the relationship that the client can keep secret (see chapter 15).

Sexual preferences are clearly indicated by Mars's position and aspects, and for women Mars also clearly shows the kind of man that is attractive and the kind of experiences had with that man. Cultural influences mean that women are encouraged to develop those sides of their nature represented by Venus—being attractive and loving—

ENERGY

Values:

Drive and action. Sexuality, desire, and wanting. The will to attain and to win. The need to compete.

Behavior:

Self-assertion. Defining limits and saying "no." Being direct and clear. Ego-centered, seeking missions and goals. Aggressiveness

Action Polarities:

Dominating/Weak. Direct/Evasive. Masterful/ Wimpy. Voracious/Controlled. Unpredictable/ Planned. Violent/Gentle. Lecherous/Disciplined.

In Society:

Iron, tools, metal instruments, machines. Mechanisms and The color red. Speed and risks. Violence and battle. Victors, mechanics. Weapons. Men: brothers, fathers. Male sexuality. sportsmen, pioneers. Doctors, especially surgeons. Engineers.

Body/Mind

Male sex organs. Muscles. Adrenalin. Blood. The front teeth. Fight or flight. Sexual drive and male identity. Courage and cowardice.

Fig. 19. Manifestation of Mars energy. Mars's energy, combined with its sign and house position and any aspects, shows the ability to manifest will and action and manifests as different styles of getting what the individual wants.

while men are encouraged to develop sides of their nature represented by Mars—being competitive and tough. (It's interesting that society spends about the same amount of money on the beauty industry as it does on armaments.) Because of these cultural stereotypes, the consulting astrologer can describe the experience of women in a man's chart through Venus, and of men in a woman's through Mars. But in the end, both sexes must own both planets equally, otherwise they will not be able to influence what is going on sexually and in love. Projection is always a form of blindness.

Now, at the beginning of the 21st century, a lot of men have an uncomfortable relationship to their Mars. Certainly the male generation born from 1940 to 1970—embarrassed by the John Wayne stereotype that blundered into Vietnam, and having gone through the sexual revolution of the 70s (when both Uranus and Pluto transited Libra)—developed a softer way of being. Though this was politically correct, it generated its own form of problems in subsequent relationships. Throughout my consultation practice, female clients have complained with regularity that their man did not give them enough resistance and allowed himself to be dominated by her in the relationship. Being comfortable with Mars means being able to draw a line in the sand, make a stand, and insist on something that is felt to be right or say no to something felt to be wrong, without caring if this meets with the other's approval or not.

Working on Mars in the consultation is a very satisfying experience because the client easily identifies its manifestation and because psychologically it's somewhat simpler transforming Mars energy than that of the other inner planets. Mars is directly related to action, behavior, and the will, and great psychological transformation does not have to be effected to make changes, though it can sometimes help. Often it is enough to describe the consequences of different kinds of Mars behavior to enlist the client into making changes. When a woman with Mars-Pluto learns that her projections ultimately make her male partner, however strong, into a wimp, then she sees it's in her interest to act differently. When a man with Mars-Saturn realizes that being a workaholic is not conducive to a good sex life, he will also be motivated for change. Mars responds very well to a training program.

Troubles connected with Mars are expressed as anger and its corollary, frustration. Further problems can be seen as sexual dysfunction and the tendency to dominate or be dominated both in personal and professional relationships. For a woman, Mars works well when she is comfortable with men but can be feminine without feeling dominated, when she can make her way in the world, and when she has healthy desires and the energy to pursue them. For a man, Mars works well when he doesn't shrink from competition, doesn't whine, and can overcome resistance in life without giving up. The virtues of decisiveness, courage, and valor are the badge of Mars.

Mars-Jupiter: Learning to Win

The combination of these two planets generally gives abundant masculine qualities, which manifest as a drive to be the best, to come out on top, and to be right. With the square, opposition, and conjunction the urge to win tends to be overly dominant, often resulting in self-righteous behavior that accentuates the need to be on top at all costs. There is little consideration for the sensitivities of others, as the overwhelming energy used to make a point or achieve a desire precludes being receptive to the response of the other person. Mars-Jupiter people can be very blunt, and they think they are doing people a favor in being so.

This will become obvious to the astrologer early in the consultation, as the client tends to act in a very knowing way to the information delivered. Though extremely eager to get at the truth, and avid in their pursuit of knowledge and wisdom, Mars-Jupiter clients have a long history of intellectual competitiveness where they falsely assume that to show ignorance is to show weakness. They have the capacity to develop a fierce wisdom, but their weakness is that they fail to rise to their full potential because of a lack of receptivity. They feel disadvantaged in the consultation because the astrologer is obviously in possession of knowledge denied to the client. Their natural reaction is to wrong-foot the astrologer, and they will be looking for evidence that the astrologer is getting the interpretation wrong. As the art of consultation is a difficult one, evidence of ignorance will not be long in coming. There may be little hesitation in informing the astrologer he or she is wrong, and remarks like "I knew that" and "You haven't told me anything new" may be delivered, making the astrologer feel very uncomfortable indeed.

When you detect this behavior, it is crucial that you do not get defensive. Putting people at a disadvantage is standard practice for the Mars-Jupiter client, and it is precisely here that the client must be confronted. Judicious questioning will reveal that brothers, fathers, or teachers browbeat these clients early in life. It is typical for the father to pontificate about strongly held convictions of a political, social, or intellectual nature. Mars-Jupiter children have had to develop equally strong ideas and convictions to hold their own in the family, and they will become used to making as strong an impression as possible to get through. There may also have been exaggerated expressions of masculinity through rough games, competition, bullying, and risk taking. Winning is what matters. Both boys and girls with this aspect will therefore develop overly forceful behavior.

A woman with this aspect will tend to be attracted to a dominating man whom she admires for his masculine and intellectual qualities. Yet, as the relationship develops, she will find she tires of the man's pontificating and need to be right all the time. She will often complain to the astrologer that he is "laying down the law," and she will tell of losing her respect for the qualities she previously admired. The man is not so clever as she thought . . . no . . . in fact, there is evidence that she actually knows more than he does; he is full of bluff and bluster. For his part, the partner cannot understand that the wise thoughts and strong opinions that were so attractive to her in the beginning now seem to antagonize her. What is happening is that the man is the same as he always was, but that the woman has absorbed ideas and opinions from him to develop her own Mars-Jupiter qualities.

People with this aspect usually have a strong sexual appetite, though there is no reason for this to be a problem, unless the partner happens to be a sensitive and romantic type and longs for the finer arts of seduction. A man with a Mars-Jupiter partner can often feel quite threatened by the woman's sexual appetite, and some men when overwhelmed in this way lose their own appetite for sex. Men who have the aspect often see themselves as something of a stud and may not find satisfaction in just one partner. Though there is plenty of moral indignation about intellectual or political issues, there may be little moral restraint in sexual areas. The randy university professor is a Mars-Jupiter archetype.

In the consultation, you must be prepared to lock horns intellectually with the client, otherwise the client loses respect. Yet there is no point in having a long drawn-out discussion—this is the atmosphere that the client thrives in. It's far better to harness the wisdom of the client into solving the problem that has manifested. If the current relationship is one of a series that has degenerated into argument and self-righteousness, what connection might it have with the client's way of being? As the client is so knowledgeable, what would he/she suggest? The consequence of a badly administered Mars-Jupiter is interminable argument. You should challenge the client to identify the root cause of this and take action.

What works for Mars-Jupiter people is a life of challenge and adventure, with plenty of competition and intellectual challenges. What is cathartic for them to admit is that they nurture feelings of arrogant superiority and believe themselves to be right and others wrong.

Mars-Saturn: The Hard Slog

People with this combination of planets are generally hard working and conscientious, and they develop great determination and will, with an amazing capacity for self-denial. It is as if Saturn is harnessed to rein in the wild horses of Mars, so passions and desires are denied fuel. While they strive for long-term goals, they have a low expectation of success. Mars-Saturn people tend to work hard rather than work smart, and they are at a loss to understand why their peers seem to attain success with such ease while years of effort on their side produce such meager results. When they meet resistance—and they often do—they doggedly dig in their heels. Yet time and again they are beaten, and this adds to their sense of fighting against the odds.

Of all planetary aspects, Mars-Saturn is the most likely to indicate actual physical violence in childhood. Recent laws in Sweden, for example, have made hitting children illegal, but it is probable that other methods of restraint are then applied that are equally cruel. Whatever the circumstances, it is likely that the child ran into some heavy-handed discipline through the father, and sometimes through elder brothers and other male authority figures. These children develop body armor at an early stage and will often psychologically detach themselves from their physical senses. The body becomes subject to an iron will, and it is often the case that Mars-Saturn people take up bodybuilding later in life. This partly reinforces the need to develop body armor, but at least it brings the benefit of being more body conscious.

This relationship to the body lies behind the sexual frustration many Mars-Saturn clients complain about. It is quite common for these clients practically to give up on sex as a relationship settles down into the long term. When asked how often they make love, they frequently say about once a month, if that. They will often claim it is because they have worked so hard and are too tired, which may well be true. The Mars-Saturn energy means that sexual arousal is a slow process, which often stops dead each time the person summons into the imagination past failure. Curiously, men with this aspect can be more in synch with the sexual rhythms of their female partner, being slow to start but also slow to finish. Nevertheless, the man does not like being a slave to his physical desire, and he prefers to keep it in control.

Women with this aspect, being slow to arousal, having mixed feelings about sexual desire anyway, and resenting being under the dominance of a man, will often find that sex is over before it has really begun. Unless she informs and instructs the man about her own slow rhythm and about what it takes to really arouse her, sexual frustration can persist for years. If she can take this step, the man would be only too delighted to follow instructions. There are, however, a number of factors that militate against changing the pattern. As a girl, it is almost certain that she experienced men—brothers, fathers, teachers—who were unfeeling and autocratic, if not downright violent. As a result, she developed fear and resentment of male authority. A woman with persistent sexual problems must first ask herself if she is allowing this resentment to taint later relations, because in the bedroom, she actually has the power to give and withhold. In the consultation, the Mars-Saturn woman will often complain that the husband is impotent or no longer expresses an interest in sex. The reason is that the male ego simply cannot tolerate sexual rejection, and rather than subject himself to it, the man would rather work more and give up in bed.

Another factor is the deadening of the body sensations that comes from wearing body armor. It is so important that there is a long ritual of foreplay, with plenty of massage and reassurance. To restore sexual satisfaction in a relationship, it is crucial that the Mars-Saturn woman is encouraging and reassuring about the male partner's attractiveness, just as she, too, needs reassurance. The Mars-Saturn man should overcome sensitivity about performance and share his fears about impotence with his partner so they can overcome problems together. There is nothing like a good manual—talking about sexual problems is the first step toward resolving them.

The challenge of difficult aspects between Mars and Saturn goes beyond personal relationships. Mars-Saturn people tend to gravitate toward jobs that seem to enslave them and that are physically exhausting. It is difficult, but certainly not impossible, for them to break the chains and go for what they really want. They are often in a position where others exercise authority over them, and this is something they really resent—and it shows. If the client complains about authority, then they are replaying an old story from childhood and using defensive techniques that only exacerbate their situation. If the story can be identified, then the present-day behavior that attracts authoritarian superiors can be changed.

What works for Mars-Saturn people is being involved in a challenging and demanding occupation that requires long hours and hard work. Ambition will always be a major factor. What is cathartic for them to admit is that they resent people with authority and are motivated more by fear of failure than desire for success.

Mars-Uranus: Shock Tactics

Contacts between Mars and Uranus give an electric, magnetic aura; astonishing dynamism; and a real ability to act under extreme pressure. It is as if these people are in the doldrums when existence is humdrum but come into their own when the pace quickens. Therefore they engineer situations when peak performance is required, while they tend to be restless and unpredictable when it is not. You get a sense that anything can happen around these people, and you can never be sure what they will do next. Their movements tend to be quite sudden, and they are always ready to strike when an opportunity presents itself. Although usually well in control, these people often contain an explosive anger under the surface, and this adds to their aura of unpredictability.

Mars-Uranus people tend to like doing things their own way, and they can scarcely bear to follow orders. In the long run, they cannot disguise this, so if they are in a difficult job or relationship, they may bear things for a while, but if it's more than they can stand, they will up and quit with no warning. At this point, it is probably too late to get them to change their mind. This dangerous impulsiveness can mean there are too many interruptions in their development for them to plot a consistent course. Their need for excitement will often mean considerable long-distance travel, particularly with the opposition aspect. They make wonderful troubleshooters, but poor administrators.

In relationships, the need for excitement means that they can make impulsive moves that can threaten stability. They are subject to sudden erotic attractions, especially when in exotic locations. In their intimate life, they are quickly aroused, which can be an advantage for women with this aspect as they often have no taste for long, drawn-out foreplay, but not for men whose partners may prefer them to slow down and take their time. They have something of a kinky streak, and like sex in unusual places where the element of spontaneity, or better still, danger, is there. As Uranus has a strongly rational and airy effect, there is a tendency to take sex outside the sphere of the physical and into the realm of the theoretical, giving a taste for experimentation.

A woman with this aspect is rarely attracted to a conventional man from her own social circle. It is very usual for her to choose a relationship where there is a great difference in race, culture, religion, and so on. In this way, the relationship always seems fresh, new, and challenging. Yet even in these circumstances boredom creeps in. The Mars-Uranus woman will often complain of violent and unpredictable behavior from her mate, but this is often due to the fact that she can be really provocative. If things get cozy and safe, she will jab at the man just to recreate the atmosphere of excitement, which is what the relationship is all about for her. Because this aspect brings a fearlessness bordering on madness, the Mars-Uranus woman can find herself in very dangerous situations.

People often say of their Mars-Uranus friends and lovers that they feel as if the latter can walk out at any moment, as if their suitcase is always packed, metaphorically speaking. This creates an atmosphere of uncertainty that is not conducive to a harmonious and loving relationship. The threat is real, for as tension increases in the Mars-Uranus person, the strategy is to get up and go, to avoid exploding on the spot. Generally, this behavior can be traced to extremely uneasy situations in childhood. One client told me what it was like to live with a father who would simply lunge out and strike her for no apparent reason other than that she probably had done something naughty earlier. For both sexes, confrontation with arbitrary authority brings out rebellious tendencies at a very early age. Mars-Uranus people are very contrary and are likely to do the opposite of what they are told and step off the tried-and-true path. They respond to authority by being provocative, as if to signal that they are not frightened of punishment, which can result in them being punished quite a lot. Other key events can be sudden upheavals or accidents, which give a permanent shock to the nervous system. It's the arbitrariness of the event that is so disorienting and that instills in the person the need to respond quickly to any emergency.

To achieve happiness in relationships, women with this combination should identify precisely what it is they say and do that drives their male partner to distraction and fury and understand the core reasons for this. If there is a complaint about unpredictable or angry partners, then it pays to reconstruct the scene so that the client's true behavior is clarified. Men with this aspect must understand that they wield power by making their partners nervous and frightened and that the threat of unpredictability is an unfair tactic.

What works for Mars-Uranus people is to have exciting and challenging work that gives them the freedom and opportunity to travel and runs counter to conventional methods. What is cathartic for them to admit is that they are needlessly provocative and choose the life of an outsider because of immature antisocial behavior.

Mars-Neptune: Desire and Sorrow

People with aspects between Mars and Neptune are often very socially motivated and want to work for the greater good. Yet they are often led astray by very strong desires that undermine their will and moral fiber. These desires may bring the person great unhappiness when young, but through a process of spiritual refinement they often develop a life of purity and restraint later. Indeed, it is the only way. They tend to be creative and have a wonderful imagination, which enables them to dream and then achieve their dreams. These people have a softness and sensitivity that makes them shun aggression and competition.

This aspect, more than any other, tends to indicate substance abuse, either personally, or among close family members. Certainly, any early experiences with drugs will open doors of perception, which lock Mars-Neptune into a spiral of desire for the ineffable and the magical. In the same way, the Mars-Neptune person may develop sexual addictions, where the desire for the perfect sexual experience becomes a Holy Grail that leads the person through a labyrinth of morally doubtful entanglements. Yet, however much the desires are satiated, there remains a gnawing dissatisfaction—"Is that it?" Mars-Neptune people are drawn to pornography and film, and while there may be a case for this kind of interest, the same dynamics apply; there must be something magical at the end of the rainbow, but they don't quite get there.

A woman with this aspect will occasionally complain that her partner abuses alcohol or drugs, and indeed she may have serial relationships where this is the case. One reason for this is that she is attracted to men who can be "saved." Another is that Mars-Neptune in a woman's chart gives such high expectations of the man that they cannot possibly be fulfilled. After a few months, the magical nature of the man—his creativity, artistic skills, musical ability, or whatever—fades, and his smelly socks and seemingly shifty nature take over. After initial adulation there is a palpable disappointment that the man has not fulfilled the dream. This disappointment can be responsible for a change in behavior later. For example, a man who knows that he disappoints his partner may turn to drink. The Mars/Neptune woman needs to identify and change the signals that have convinced her mate he is a failure in her eyes, rather than focus on his drinking. The latent substance abuser is in reality the Mars-Neptune client, and analysis of the dynamics of the relationship often shows that they both indulge in an attempt to recreate the early magic.

Another frequent complaint is infidelity. Surprisingly often, the Mars-Neptune woman discovers that a man has been unfaithful to her for years without her knowing. One of the factors that accounts for this is sexual. Mars-Neptune hates routine sex and loves seduction. But even seduction becomes routine in a stable partnership. During sexual relations, the Mars-Neptune woman does a number of things that in the long run can contribute to her partner's infidelity. She may have strong fantasies from which the man is excluded. She should share them, otherwise the man unconsciously feels alienated. The Mars-Neptune client will also be very considerate during lovemaking, often agreeing to have sex when she does not really want it, thus losing the magic. She must be firm about saying when she wants sex and when she doesn't.

It is often the case that the major male figures in childhood exhibit some kind of basic weakness, which undermines faith in the masculinity and manhood of the Mars-Neptune person. Quite often, the father shows weakness, usually in relation to an apparently dominating wife. The father shares his tears and sorrow with the child. There can also be a slightly inappropriate sharing of personal difficulties between father and child, even of sexual frustrations. Mars-Neptune becomes used to and expects to see this weakness later on. This accounts for the common complaint of the female Mars-Neptune client that her partner is wimpy. First she is attracted to a sensitive man; later she evokes weakness by focusing on him in a perfectionist way.

For the man with Mars-Neptune there is far less projection; the drama takes place within. Apart from struggling with moral and sexual issues that threaten to undermine any progress, he also finds it hard to come to terms with a man's world that requires an aggressive and competitive attitude. The very idea of routine work awakens dread, and professional happiness is elusive unless the man finds an area in which he can pursue his ideals or his creativity. The advertising and media industry is an ideal playing field, especially because they deal in intangibles that may or may not have real worth. There are also real healing talents with Mars-Neptune and, when harnessed politically, a caring social conscience.

In the consultation, it pays to identify issues concerning sex and deception, addictive behavior and moral dilemmas, especially when the client seems to be the victim of *others* in this respect. Somewhere, the client is doing something that evokes this kind of unwanted result, and the behavior—when identified—can be changed. Unrealistic dreams need to be purged: there is no perfect man, sex will never consistently transport you to heaven, and there is no path to success without hard work and dedication.

What works for Mars-Neptune people is to have a creative job that corresponds to their ideals and to have a partner who can share in imaginative ventures. What is cathartic for them to confront is the inner sleaze, the secret immoral imaginings, the subjection of will to desire.

Mars-Pluto: Power and Powerlessness

People with this combination of planets have an enormous impact on others, but are often unaware of it. They respect strength and abhor weakness, and they have the power to subdue fear and engage in long battles to transform situations that they believe need reform. Inwardly, they are subject to many fears and phobias; outwardly, they appear imposing and rather frightening. They exude an atmosphere of control and a dark magnetism that puts people under their spell.

Mars-Pluto clients have no respect for a hesitant and timid astrologer, and they will soon establish control if this is the case, cutting themselves off from the possibility of benefiting from the consultation. It is necessary to overcome trepidation and make bold statements, even at the risk of being wrong, rather than tentatively trying to extract information. Naturally suspicious, these clients want hard evidence that you know your stuff. Force must be matched with force. As Mars-Pluto clients have very deep anxieties, establishing control is a way for them to feel safe, and in the one-to-one consultation they will try to impose their will. Inexperienced astrologers can go through the consultation worriedly trying to prove their worth and end up feeling completely exhausted. The Mars-Pluto client leaves without his cover being broken, but also without having made any progress.

The way around this is to accurately describe the manifestation of this aspect and its consequences, because they are so extreme that the Mars-Pluto person will easily identify them and be powerfully motivated for change. Mars-Pluto people are terrified of losing control and expend so much energy maintaining it that they exhaust themselves and feel at the mercy of unpredictable developments. To get by in life you have to trust: that the elevator cable will not snap, that you will not be mugged, that people will not rip you off, that the taxi driver will not crash. The Mars-Pluto person tends to organize life so that as few things as possible have to be trusted, and a lot of energy is expended on this kind of surveillance and control.

The woman with difficult Mars-Pluto aspects tells the same story again and again. She is attracted to mysterious and powerful men who seem to offer some kind of protection in life. As the relationship progresses, she is riveted to the weaknesses and flaws that reveal themselves in his nature. She instructs him how to improve, so that she can be less nervous about him. It's a Catch-22 situation. If he changes, she will perceive him as weak and lose faith. If he doesn't change, she will be angry with him and wear him down psychologically. (And the advice to the partner who undergoes this is: resist change—maintain your self-respect!) The result is that after some time, the woman loses respect for him. She doesn't just see him as weak, but as pathetic. He doesn't just beg to get into her good books, he crawls (and she hates him for it). Alternatively, the man really is strong but also dictatorial, and the relationship degenerates into a long series of humiliations.

Fear of not having control and being under the power of another also affects the sexual life. Men with this aspect are driven by sex. A deep-seated anxiety about their virility paradoxically drives them to demand sex all the time. It is not unusual for a man with a strong Mars-Pluto aspect to have sex three times a day and to imagine that his stud-like behavior is the answer to his partner's dreams. In my practice, I have on several occasions had both the man with the Mars-Pluto aspect and subsequently his female partner as client. The woman perceived her partner's sexual demands as intrusive and exhausting. More importantly, she pointed out that her partner's potency, which was not in doubt, manifested as a kind of robotlike sexuality, which made her feel like she was making love to a machine.

The woman with this aspect also has deep sexual needs but has difficulty surrendering to them. She will often initiate sex, and if it does not go according to plan, she will instruct her partner as to precisely what to do. This is again a Catch-22 situation because if he allows himself to be led, sex loses its magic for her. Sex might involve some kind of humiliation or games that involve control. The root issue here is that to have an orgasm there must be surrender, and surrender is what Mars-Pluto fears most.

Along with an obsession with sexuality comes its corollary—jealousy. The Mars-Pluto man, in particular, is ridden with jealousy, imagining that all other men have the same kind of sexual designs as he has. This jealousy is completely concealed, for to reveal it would be to show weakness. In extreme cases, the man will shadow his partner or hire someone to do it. At the same time, the man will have extremely secretive behavior, and his partner will be at a loss to find out what he is doing with his time. The woman with this aspect will often claim that her male partner is jealous, and again, in extreme cases she will be shadowed by the man—even though it is not him that has the Mars-Pluto aspect! The reason for this is threefold: first, she does not share what she is doing with the man, keeps her actions secret, and therefore arouses his suspicions. Second, Mars-Pluto in a woman's chart, just like Venus-Pluto in a man's, does not respect the natural space that exists as an aura around people. She comes close in to men, establishes physical contact, and indeed has a deep psychological understanding of them. She engages in inappropriate intimacy that drives her partner to distraction. Finally, a jealous man is a man who can be controlled.

Where does this need for control come from? In childhood, there may have been extreme situations relating to the father, and brothers (if there were any) and possibly other key male figures. There are often half-brothers in the background and jealousy connected with them. Sometimes there is a brother who terrorizes the child, and often there are episodes that infringe sexual boundaries. These episodes can happen at an early stage, especially concerned with potty-training, where there is a battle of wills and subsequent humiliation. At the dinner table, too, there tend to be drawn out battles where the parent tries to assert his or her will for one reason or another. Whatever the situations were, they tended to awaken deep anxieties, and the subsequent behavior of the Mars-/Pluto person is simply geared to allay these anxieties. They resolve never to let another person have control over them, and this means that they are compelled to control other people.

In their professional life these people often get entangled with faceless, all-powerful authorities such as corporations, unions, the police, and tax authorities, as if their worst nightmares regarding powerlessness are fated to come true. They can sometimes effectuate reform in these institutions, but just as often they are crushed by them. The problem is that people with power are demonized—and the result is that the Mars-Pluto person ends up fighting a losing battle against superior forces.

The energy of Mars-Pluto is very transformative; there is a good chance of completely changing the behavior associated with it in the consultation. When the client realizes that anxiety has genuine roots in real events and that the effort to exert control is counterproductive and based on inner fears, then they want to change. They are horrified when they realize the effect their behavior has on others, particularly in intimate relationships, and often spontaneously reform. They are often drawn to psychotherapy and can benefit from it.

What works for Mars-Pluto people is to exercise real power in work that requires an element of ruthlessness or investigative abilities. What is cathartic for them to admit is that they are manipulative and play power games, re-enacting primitive survival scenarios without concern for the sensitivities of others.

Meghan

Jupiter through Pluto—The Collective Framework 9

More often than not, the behavior associated with planetary aspects is based on complex aspect patterns, rather than the interchange of energy between just two planets. There are innumerable combinations, each one generating unique behavior, but of greatest significance is the combination of a personal planet with two outer planets. Outer planet combinations show long-term, collective trends that affect whole generations and specific groups within them. At any given date of birth there are vast social and political changes sweeping through society, embodying a particular spirit of the times. A person born on this date will embody the aspirations of that time, and these aspirations will become crucially important at a time in the person's life when the planet combination is repeated. For example, people with Jupiter-Saturn aspects will find key moments arising every 20 years (and also at the square and opposition points of the cycle—every five and ten years), while people with Jupiter-Pluto aspects will find key moments arising every 13 years, and those with Uranus-Saturn aspects every 46 years, and so on.

The ten major planetary cycles have a unique signature, and when a personal planet interacts with an outer planet pair, the astrologer should be able to identify a specific corresponding behavioral pattern. The following descriptions relate more to behavioral patterns than the social and political trends. As in the earlier section on aspects between two planets, the following texts concentrate on the difficulties that may be apparent in the consultation, rather than on the many positive manifestations of the triple-planet aspect. The descriptions here are aimed at those clients who obviously manifest the more extreme problems associated with the most difficult combinations.

Jupiter-Saturn: Structures in Society

Those born with aspects relating to the two largest planets in the solar system are strongly keyed into mainstream society in one way or another and have the capacity to make visible changes in its structure. With the positive aspects in particular there is a gift for being in the right place at the right time and for fulfilling the needs of an evolving society. Those born with the conjunction have the potential to become pillars of society. With the square and the opposition, the need to achieve something is profound, but there is less confidence, as the faith of Jupiter meets the obstacles of Saturn. In the consultation, the relevant areas to examine are work, goals, and ambition, as well as the relation of the individual to society, with a view to maximize the client's contribution in this area.

Sun to Jupiter-Saturn

These people have the capacity to be true leaders in their field. Generally, they will have an aura of responsibility, and their opinions will carry weight. The father usually is influential and rather dominating. These clients take themselves seriously and are eager to make a mark. They can see setbacks in a positive light as gaining experience.

Moon to Jupiter-Saturn

Generally, these people are serious by nature, often quite hardened by early events in their life. They have the capacity to see the posi-

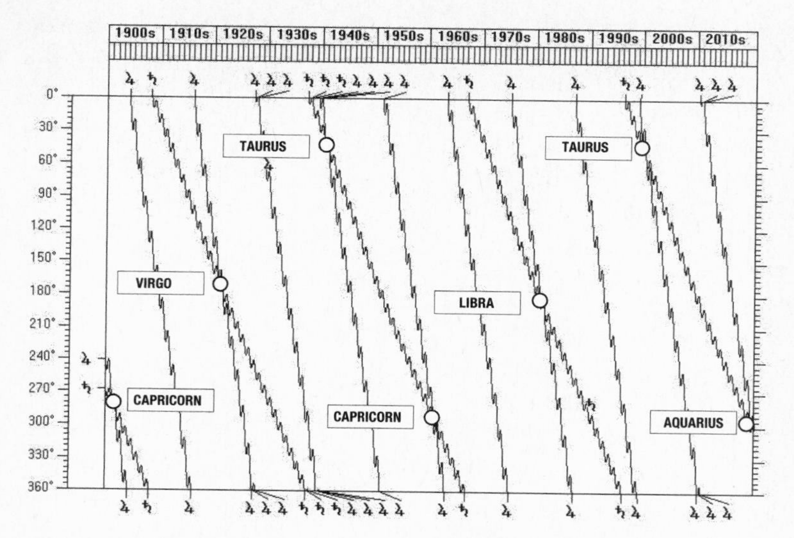

Fig. 20. Jupiter-Saturn conjunctions 1900–2020. These conjunctions take place every 20 years. In 2020, there is a transition from a series of conjunctions in earth signs to a series of conjunctions in air signs.

tive side of negative events, after initially being weighed down by them. Family responsibilities are a major issue, as well as an anchor in their life.

Mercury to Jupiter-Saturn

This is the signature of a mental achiever, probably powerfully motivated by parents or siblings in childhood. While early intellectual achievement may have been characterized by setback, these people will often go on to make significant contributions to society in areas such as education, communication, or transport. It's a great combination for business success if the client overcomes negative thinking.

Venus to Jupiter-Saturn

The need for a stable relationship features strongly here, as well as an urge to advance socially through marriage, partly because of a desire for tangible proof of self-worth. This type will have a strong focus on finances and the probable ability to increase assets by judicious investment. This is an excellent combination for business, and both material and social advancement will feature as values when the client chooses a partner.

Mars to Jupiter-Saturn

This is a tough-guy combination and, if Jupiter is dominating, somewhat of a bully. Issues of self-assertion dominate, and both sexes have the ability to get by in a man's world. Dynamic and hardworking, these people advance through both effort and confidence. There is a determination to win and great sensitivity to defeat. Physical and intellectual prowess are highly valued.

Jupiter-Uranus: Creating the Future

This combination is a "New Age" influence par excellence; those with the aspect tend to have an international perspective and wide horizons. Travel to exotic destinations and/or interest in unconventional subjects is a feature of their lives. They have a dominant belief in the freedom of the individual and a pronounced aversion to any authoritarian institutions that limit physical or intellectual freedom. There is generally a talent for, and interest in, new technology among these people, and they remain young revolutionaries for their whole life.

Sun to Jupiter-Uranus

Influenced by a perception of the father as free-thinking, unreliable, or unconventional, these types strive to be enlightened in their dealings with others, setting an example for others through a warm embrace of libertarian principles. They are detached in style and sometimes remote. Fearless in expressing opinions that run against the norm, they nevertheless tend to value friendships as extremely

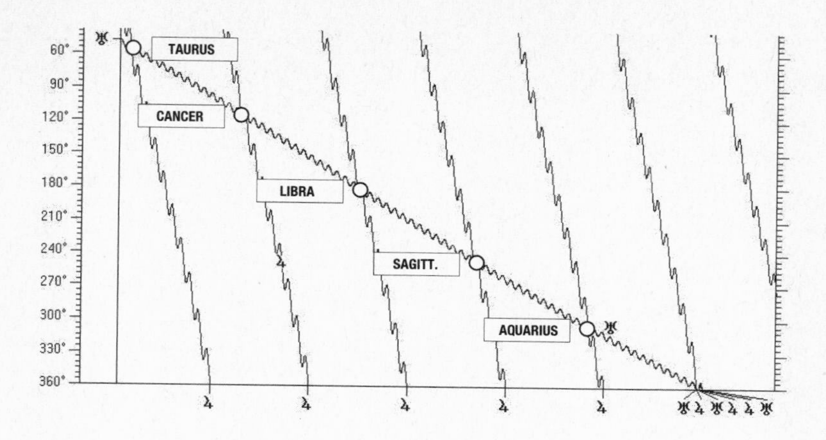

Fig. 21. Jupiter-Uranus conjunctions 1940–2010. Taking place every 13 years, these conjunctions herald new technological advances that change the spirit of learning, depending on which sign they occur in.

important. They take a stand for change in society. Although open and tolerant, they have a tendency to be immovable in their opinions.

Moon to Jupiter-Uranus

Usually characterized by an unconventional and unstable family environment or a very untraditional mother, these people will often have strong connections with foreign cultures and even live abroad. They love freedom and are indifferent to security, but inside, their feelings are restless and unstable, making traditional relationships rather changeable. They have a habit of precipitating emotional upsets and then reacting remotely and intellectually to the resulting storm. As these types appear to be cool, reasonable, and enlightened, their responsibility for the disturbed environment is often obscured.

Mercury to Jupiter-Uranus

There is always an element of genius with this combination, in one area or another. These people have an interest in and natural talent for new age subjects such as astrology. Generally their mind is lightening fast, and very intuitive. It is easy for these types to get bored, so they are not particularly suited to long-term educational projects. They are best at gathering learning in bursts and function best when under pressure with limited time. Under extreme circumstances, they may suffer from mental disturbance, as if their thought processes were short-circuited.

Venus to Jupiter-Uranus

People with this aspect have predominant values regarding the principle of personal freedom. This is a "feminist" influence, but neither sex is willing to subject themselves to limiting or conventional love relationships. It is highly likely that they will at some point have a relationship with an exotic stranger, foreigner, or culturally challenging person. This person gets along far better in groups than in a monogamous relationship and is prone to sudden romantic attractions, especially if they can learn something important from them. Behind an espousal of "enlightened" principles of mutual freedom in a relationship lies a fear of the perceived inertia of commitment.

Mars to Jupiter-Uranus

Principles of personal freedom reach their most powerful with this planetary combination. Even the slightest hint of control by "authority" is perceived as intrusive, and this person will never let slip the opportunity to demonstrate complete independence. This person is likely to have a very unusual career and considerable original talent in the chosen area. Generally, these people are fearless, if not reckless. It is not in the nature of these types to be sexually monogamous, because their need for sexual freedom and experimentation is so strong.

Jupiter-Neptune: Visionary Dreams

The unique signature of this combination is a utopian idealism and incredibly vivid imagination. It is very much related to film and image and to dreams or fantasies that grip the imagination of society. In the individual, visions will often founder on the rocks of reality, yet succor is sought and found in an inner dream world. These people have the ability to understand the power of the unreal, whether it be through advertising, brand image, or any other method that channels the escapist aspirations of the public.

Sun to Jupiter-Neptune

These people are true visionaries who have disdain for the constraints of the real world and are strongly motivated to lead others

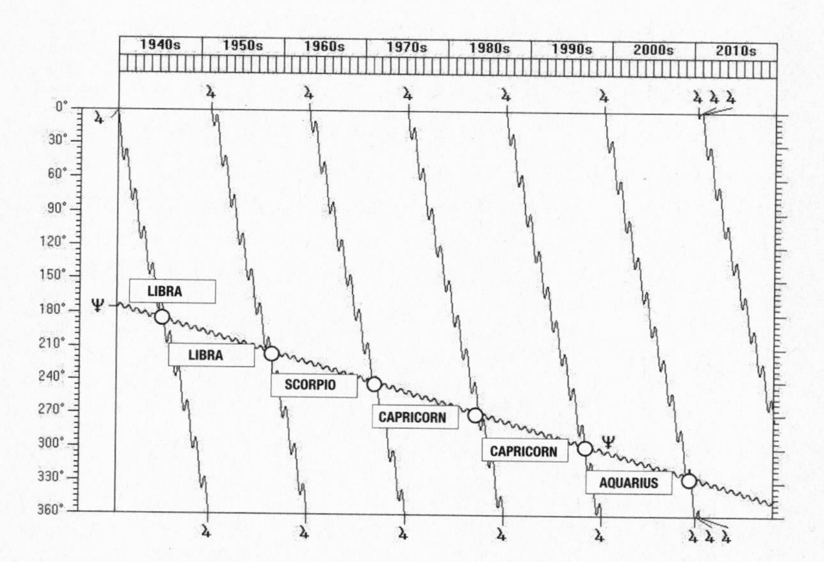

Fig. 22. Jupiter-Neptune conjunctions 1940–2010. Occurring every 13 years, these conjunctions herald a euphoria connected with the sign it falls in, often resulting in the bubble bursting. These are times when the media find new tools for swaying the masses.

toward a vision of the future. Depending on the aspect combination, moral restraints do not figure highly in these people's consciousness, although they can have a genuine compassionate vision for humanity as a whole. It is likely that there is an inflated and unreal perception of the father and, indeed, of themselves. The tendency to see themselves as victims of a gray and unfeeling society is pronounced; they need to weave their dreams into the fabric of reality to make a difference.

Moon to Jupiter-Neptune

Extremely sensitive and receptive, these people have the capacity to feel profound sympathy and compassion, as long as they do not fall into the victim trap—a risk that usually stems from a perception of the mother as unjustly wronged. With this combination, the emotions threaten to overwhelm the person, and he or she has to learn strategies to gain greater perspective and calm. These people generally have a talent for film and photography.

Mercury to Jupiter-Neptune

The challenge here is to assimilate the lessons of both the conscious and the unconscious mind to gain greater insight into life. This combination produces the greatest imaginative power, although how it is channeled depends on the moral values of the individual. These people are sorely tempted to tell "whoppers." One problem these individuals face is a difficulty in distinguishing fact and fantasy and in being real. They tend to believe what they want to believe and say what they think needs to be heard. Even when telling an untruth, they can truly convince themselves that it is truth.

Venus to Jupiter-Neptune

With this combination, the person struggles with the dream of the big romance and a corresponding inability to live up to the everyday demands of a normal relationship. These people have a tendency to idealize their partner, especially in terms of cultural and creative prowess. Materially, too, they dream of the big win. Generally then, their problems lie in dreaming about how things could be instead of taking satisfaction in the way things actually are. They have a gushing ebullience that gives them a sirenlike quality. Both sexes are prone to the allure of seduction, for that's when they get closest to the dream.

Mars to Jupiter-Neptune

The fiery enthusiasm of this combination can easily be diverted along morally questionable avenues. These people do have the power to effect social reform along idealistic lines or to work extremely creatively in areas of fantasy and imagination. However, more often than not, they suffer because of their sex drive, which seems to take control of their character and will and drive them to the verge of corruption. Women fall for charlatans, while men are drawn into pornographic cul-de-sacs. In the long run, they will gain insight into the emptiness of sexual desire.

Jupiter-Pluto: Ideological Extremism

The main negative signature of this combination is an idealistic fanaticism that makes these types intellectually dominating. Often they will have an attraction to cultlike groups or educational, political, or religious movements in which totalitarian ideals are expressed. This does not have to be obvious—many a philosophical group reflects this tendency. They can be recognized by the preaching of ideas that suggest that if you do not believe in them, then you miss out on the real truth. Any organization that strives to police the mind has this Jupiter-Pluto quality. People with this aspect tend to embrace a particular belief very intensely, only to drop it completely later on.

Sun to Jupiter-Pluto

Powerful and charismatic individuals, these people almost invariably experience their fathers as dominating, especially intellectually,

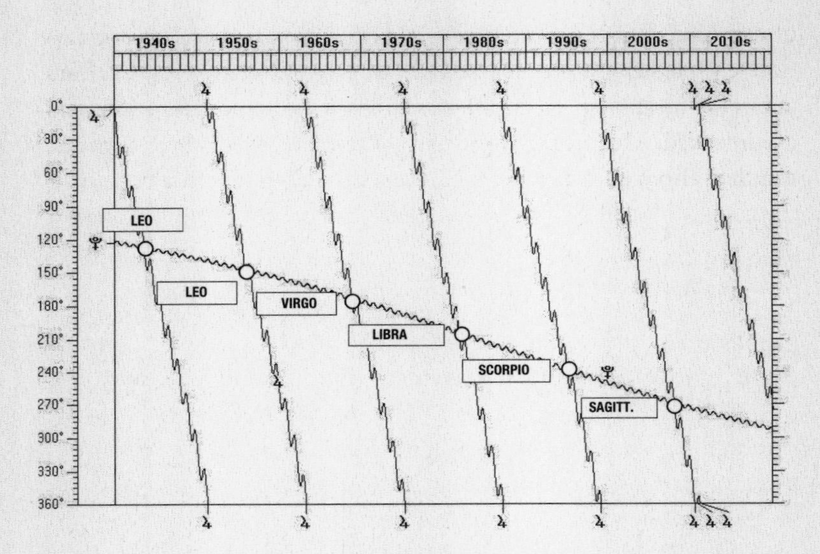

Fig. 23. Jupiter-Pluto conjunctions 1940–2010. Taking place every 13 years, these conjunctions are connected with powerful new ideological influences that transform conventional wisdom, as well as financial cartels and other small but disproportionately powerful groups.

or the father may not have been present during childhood. It becomes extremely important that they are listened to, understood, and respected for their beliefs. They achieve this by generating an intensity that tends to overwhelm most others, who can handle them only in small doses. To question, ignore, or reject their opinions is almost impossible, so they tend to get their way, yet feel alienated in the process. Their biggest life crisis in adulthood comes when a belief system they have embraced for years collapses, leaving their identity in ruins.

Moon to Jupiter-Pluto

The emotional emanations from these individuals are powerful and overwhelming, making them rather exhausting to be with in the long term. They experience their mother as dominating, probably because of dogmatic beliefs or moral admonitions. Their overactive imagination tends to exaggerate their anxiety, and they can have very real fears about the future, especially as far as children and other family matters are involved. They tend to gravitate toward extreme environments, but they also have the capacity to mobilize ecological reform.

Mercury to Jupiter-Pluto

The trouble spot here is that these types can be extremely opinionated and intense. Traumatic early experiences with friends and siblings or at school will drive them to develop their mind so that they can prove their intellectual strength. Debate for them is a matter of life or death, which means that they can be very tiring to have discussions with. At the heart of their mental intensity lies an anxiety about mental capacity. They are also in danger of doggedly hanging on to fixed beliefs at a time when abandonment of their convictions would be liberating for them.

Venus to Jupiter-Pluto

With this combination, the power of the feminine is exalted to a symbol of iconic power. The men tend to be attracted to influential and charismatic women who embody both intellectual and erotic qualities; the women with these aspects tend to be strongly aware of the power afforded to them by their sex. Both are inclined to break taboos in relationships and to go to extremes to find the wisdom therein. Often the power of money is a key factor in their life, although these individuals will be willing to gamble everything in extreme situations. Economically and romantically, it's a roller-coaster life.

Mars to Jupiter-Pluto

Few aspect combinations are as powerful as this, and it is important that these people use the awesome power available to achieve beneficial results. If they don't use the energies this way, then they develop a tendency to bully and browbeat. Men with this combination tend to have an insatiable sexual appetite, although they can often experience dramatic reform as a result of psychological crisis. The women tend to be attracted to dominating and influential partners, whom they later "discover" to be wimpy or inferior. Intellectually, there is a tendency to make mincemeat of the opponent in a gladiatorial contest.

Saturn-Uranus: Bridging the Gap

This is one of the slower-moving planetary cycles, and both the 12th-year and 23rd-year square and opposition phases are significant. The potential for this combination is to be a bridge to the future, although individuals born with the hard aspects are more often torn between past tradition and future change and find themselves identifying with one or the other. Caught between the air and the earth, they too often find their visions crashing against a wall of inertia. Colossal political change is generally taking place during these periods, and political decisions of one kind or another often assume considerable importance in the life of these individuals.

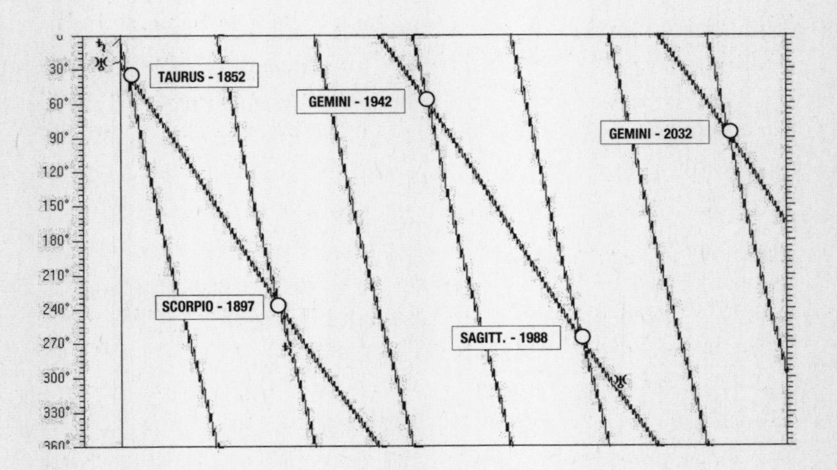

Fig. 24. Saturn-Uranus Conjunctions 1850–2050. Occurring every 46 years, these conjunctions herald revolutionary new methods that radically affect the fabric of society and international politics.

Sun to Saturn-Uranus

Powerful and independent in maturity, these people often straddle important positions in which they use their experience and power to help break through the resistance of the old guard and make significant changes. Earlier in their life, they are torn between driving ambition and a lack of confidence in their abilities, before they find their true niche. This can make them appear gauche and clumsy. Although they can embrace revolutionary ideals, they are reluctant to break their connection with the "powers that be." They have a strong willpower and inflexibility in adversity.

Moon to Saturn-Uranus

With a strong emotional character developed through extremely tough circumstances in childhood, these people tend to be inhibited, resilient, and uncompromising. While they can put up with considerable difficulties in the environment, they often entertain the idea of drastically liberating themselves from the yoke of dependency, though they would in reality be very reluctant to do this. Strong forces are at war within them: freedom versus stability; experimentation versus conventional norms; future hopes versus past events. They need to develop an inner confidence based on their achievement, which enables them to take calculated risks for the future.

Mercury to Saturn-Uranus

In a more positive manifestation, this combination shows the ability to bring innovations into the mainstream, but more difficult aspect variations put the mind under a lot of pressure. Different internal factions vie for control, as the person's need for traditional education is set up against the urge to go off on a tangent pursuing unique personal interests. These people tend to communicate forcefully and authoritatively, and they have great resilience when unforeseen circumstances occur. Often life's circumstances see these types embracing very conventional activities on the one hand and bohemian interests on the other, with neither side meeting.

Venus to Saturn-Uranus

Manifesting strongly in relationships, these people have a tendency to experience long-term commitment as restrictive and not in accordance with romantic needs. While the circumstances around partnership are sometimes rather oppressive—generally through an unusual choice of partner—these people often entertain the idea of separation as an option, creating stress and insecurity. Materially, too, they perceive economic realities as restricting personal freedom. Although many difficulties would be alleviated by a spirit of compromise, this does not come easily.

Mars to Saturn-Uranus

Willful, powerful, immovable, and incredibly resilient, these types fight their inhibitions to carve their way ahead in society. However, they do have a tendency to use too much force, or to act in inappropriate ways in the urge to burst constraints. Yet these constraints are often self-imposed. Although they have an extraordinary amount of energy at their disposal, they tend to bottle it up. In worst-case scenarios, the threat of violence overshadows their life, and they need to put up with extreme circumstances. They experience sexual tensions, too, as powerful inhibitions battle with the need for freedom and experimentation.

Saturn-Neptune: The Yoke of Time

This combination shows spirit crucified on the cross of matter, and people with this aspect will often experience suffering relevant to the house positions of these planets. They feel a sense of helplessness in the face of circumstance, and, indeed, with this aspect, surrender and acceptance are often the best courses of action. The lesson is that some

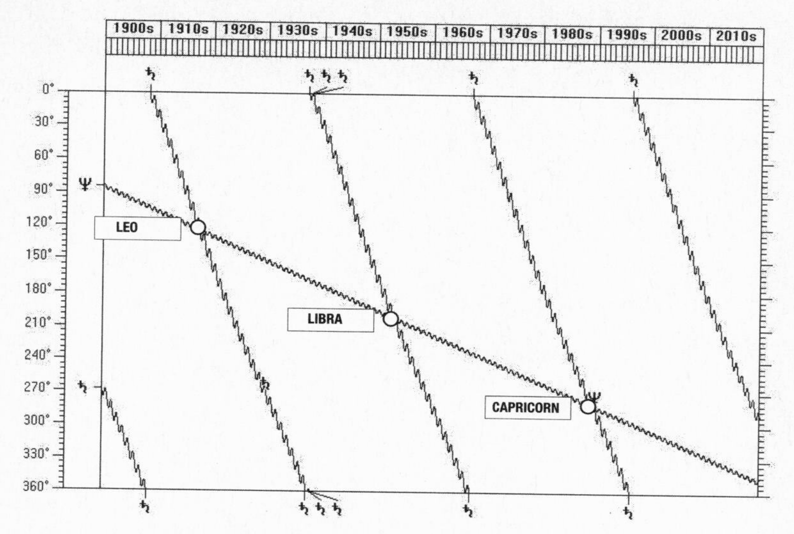

Fig. 25. Saturn-Neptune conjunctions 1900–2020. Taking place every 35 years, these conjunctions lead to the dissolution of rigid structures and a rebirth of social ideals regarding the human condition.

things in life cannot be changed and that by willingly accepting burdens there is long-term spiritual liberation. Quite often, these people find succor in dedicating themselves to others who suffer in society.

Sun to Saturn-Neptune

These people often have a long and sad story to tell about paternal circumstances, or they may simply perceive the father as deeply religious with a tendency toward self-denial. The rather oppressive atmosphere in childhood often gives these people tremendous depth of character and an ability to deal with difficult issues very sensitively and responsibly. However, long experience tends to induce a sense of negativity that undermines confidence and self-esteem. They have an air of resignation and low expectations. Transformation comes when they take responsibility for their life circumstances at the deepest spiritual level.

Moon to Saturn-Neptune

In childhood, these people may have experienced sadness and even deprivation, and their experience of the mother may be tinged with a perception of martyrdom or slavery. The profound effect of the heavy childhood atmosphere induces a depressive state that can be difficult to shake off. Old habits of duty and sacrifice tend to dominate adult behavior—one set of prison walls is replaced by another—and the tendency to accept one's lot in life in a spirit of resignation will often mean accepting negative circumstances that can, and should, be changed.

Mercury to Saturn-Neptune

Early learning difficulties and the experience of having been eclipsed by siblings or classmates tend to induce in people with this aspect combination a lack of faith in abilities, which further inhibits communication and learning. The experience of not living up to educational expectations leads to evasive mental patterns and communication, which has the purpose of disguising or obscuring flaws in knowledge. These evasive tactics will often be perceived as dishonest, and they only serve to render the whole process of communication more difficult than is necessary. Mercury-Saturn-Neptune people need to redefine their perception of failure and begin to appreciate the successes they have attained against the odds.

Venus to Saturn-Neptune

The burdens and sadness characteristic of this outer planet pair are here transposed onto relationships and the love life. Early setbacks connected with a perception of the parents as unhappy in their relationship, or with personally feeling unloved or undervalued as a child give a low expectancy of happiness. Sacrifice or oppressive circumstances may characterize later relationships—often because they feel they deserve nothing more—but it is also possible that these people voluntarily shoulder a difficult relationship burden and thereby attain redemption. Often unhappy in existing relationships, they may attempt to find happiness through affairs, but these inevitably lead to failure and disappointment. By accepting inner responsibility and embracing a compassionate approach, these people can achieve satisfaction.

Mars to Saturn-Neptune

On the positive side, this combination can portray tireless reformers who are so moved by the plight of others that they work selflessly to alleviate their suffering. On a personal level, however, people with this planetary combination have many inhibitions to overcome, especially sexually. They have a strong tendency to self-denial and even celibacy. In any event, sexual confidence and erotic pleasure are difficult for them to attain. The women tend to project their inhibitions onto a supposedly depressive and inhibited male partner; the men despair at living up to supposed ideals of masculinity. Openness and instruction can solve many of these problems. In professional life, they have a sense of long struggle without due reward.

Saturn-Pluto: Implacable Power

The signature for this planetary pair is overwhelming power, coupled strongly with fear and anxiety. It is an extremely difficult energy to handle because it keys in to primitive and atavistic fears relating to survival. It seems that there is no possibility of compromise when this energy is activated, so those who come into contact with these types will tend to develop a fear of them and end up trying to escape their domination. It seems that Saturn-Pluto people gravitate toward battles with the powers that be: the police, the tax authorities, and other shadowy groups, projecting onto and evoking in these organizations monstrous characteristics. They have an ability to handle power, but they need to do it with psychological awareness. Many find liberation when they actively embrace power, especially through shamanistic activity.

Sun to Saturn-Pluto

Generally, Sun-Saturn-Pluto people perceive the father as autocratic (in extreme cases, verging on sadistic), and much of childhood is spent mobilizing the will to withstand imagined or real assaults on the person's integrity of identity. If the father disappeared, then they may battle with ersatz authority figures. These people have a real fear of disappearing into insignificance and a corresponding compensation to manifest the Self. They generally do this with a disproportionate amount of energy, and others become overwhelmed and fearful. They can have a sense of alienation, possibly because of imagined (or evoked) mistreatment on the part of authorities.

Moon to Saturn-Pluto

No planetary combination is as atavistic as this one, the energies of the three planets keying in to deep, primitive residues in society on

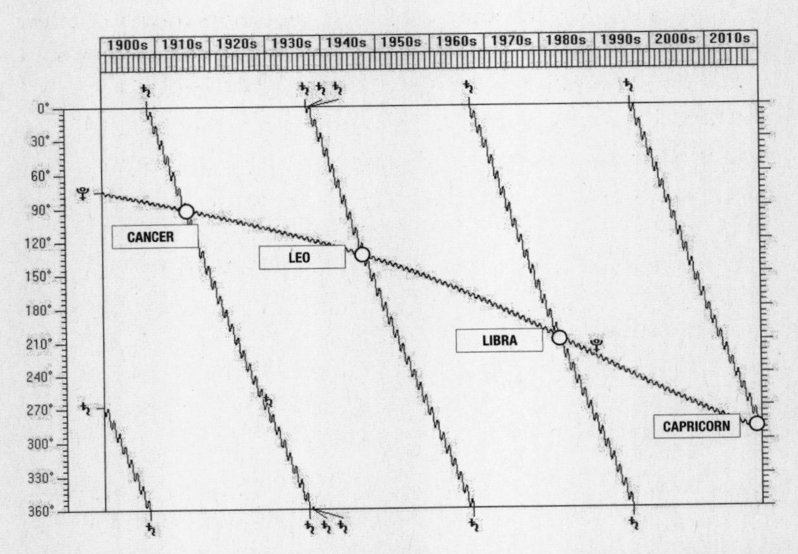

Fig. 26. Saturn-Pluto conjunctions 1900–2020. These conjunctions take place approximately every 35 years, bringing vast shifts of power in basic political structures in society, often with the threat of serious ultimatums.

a purely instinctual level. With extremely difficult circumstances affecting childhood and a mother who is often perceived as depressive to the point of madness, people with this combination have a sense of having endured great cruelty. An interest in psychology and shamanism can liberate these people from self-destructive emotional habits, which dog early relationships. No extremes will daunt these people—they can put up with chillingly oppressive circumstances. This often means that their partners go through a nightmarish and exhausting emotional maelstrom in order to assuage the anxiety felt by the person with this planetary combination. A difficult destiny passed down on the mother's side will often be brought to an end during this person's lifetime.

Mercury to Saturn-Pluto

There is a stubborn relentlessness to the mental nature of a person with this combination. Childhood factors influenced these people to learn to cope with oppressive communication methods, from repressed family secrets and taboos to mental training regimes bordering on the sadistic. Sometimes they cultivate a particular kind of obtuseness as a kind of revenge on the intellectual oppressor. They harbor an acute anxiety about mental abilities, and occasionally have some very real disability to overcome. Anxiety about mental blockage can lead to a peculiar doggedness in communication, which grinds away at facts and information. This gives great investigative ability. If the person gains a position of bureaucratic power, they may develop a taste for dominance.

Venus to Saturn-Pluto

People with this combination experience events that undermine self-worth to such an extent that it can border on the self-destructive. They have a tendency to begrudge the self and others pleasures, and have a corresponding resilience to hardship and difficulty. Relations with the mother or, more likely, a sister who appeared to have a behavioral problem lead to a fear of intimacy. Control and self-protection dominate in their choices of partner. For Venus-Saturn-Pluto women there is a tendency for early relationships to involve extreme circumstances and an element of humiliation, both emotionally and sexually. The lack of self-love that men with this combination have tends to be taken out on the female partner, resulting in subservient behavior on her part, which the man subsequently despises. Deep, transformative experiences lead to an appreciation of self-worth.

Mars to Saturn-Pluto

No planetary combination has the sheer strength and willpower of this one. Childhood events that called upon these people's ultimate survival abilities over long periods lead to the development of a dogged determination to get through whatever life throws them coupled with a low expectancy of easy success. These people gravitate to extreme jobs that require resilience, often in the corridors of power or in trade unions, locking horns in exhausting power battles, questionable victories, and humiliating defeats. They feel that to compromise is to show weakness. Women with this combination gravitate to inhibited yet dominating men, whom they tend to drive away as their own dominating nature slowly becomes apparent. The sex life is characterized by coercion. The men are obsessive about their sexual capabilities, secretly fearing impotence but often performing with a robotic staying power. This combination shows the greatest shamanic abilities.

Uranus-Neptune: Revolutionary Vision

These enormous planets conjoin every 172 years in slow, forward motion through the zodiac, so generally a set of two conjunctions takes place in the same sign over a 344 year period. There was, for example, a conjunction in Sagittarius both in 1479 and 1649—probably connected with the Reformation and the revolutionary religious events surrounding the onset of Protestantism, as well as the explosion of exploration that saw the beginning of world colonization by Europe. And more recently there have been two conjunctions in Capricorn, the first in 1821 and the most recent in 1993, which is partly connected to the development of political systems such as capitalism and communism and ideals about government systems. The Uranus-Neptune synod is connected with revolutionary vision and the power of ideas to change the human condition.

Any combination of these two planets attunes the individual to invention and developments in society that form the zeitgeist of the time. The square aspect, which took place five times between 1954 and 1956, had much to do with the onset of the effect of television culture. The imagination of the populace was held in the grip of the new immediacy of broadcasting, which generated col-

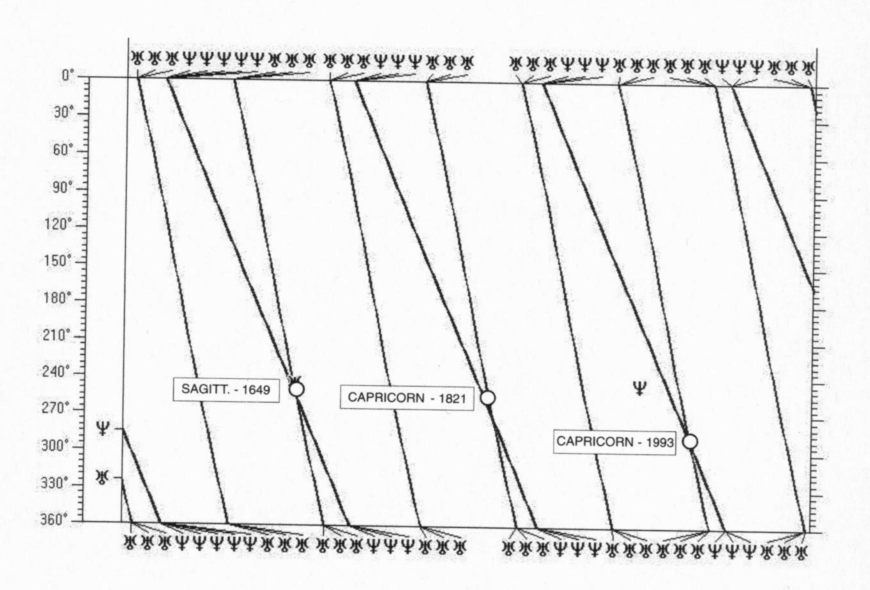

Fig. 27. Uranus-Neptune conjunctions 1600–2000. Taking place every 172 years, these conjunctions bring spiritual turning points in society, generally through revolutionary developments that change consciousness of time and space.

lective fantasies that both enlightened and misled. For example, propaganda generation reached new heights at this time, as communist and capitalist systems enrolled their citizens in convincing visions of the imaginary dangers that each side represented. So it is with clients who have this pattern that they become inspired by ideas and visions, which, though often praiseworthy, are not realistic. This may not matter if they have the ability to represent these visions creatively, because they will be able to capture and channel the imagination of their contemporaries.

Sun to Uranus-Neptune

Those born with the Sun in stressful aspect to this combination will be visionaries in one way or another. Many travel to distant lands. Contact with radically different cultures, where there often is a strong spiritual element, will awaken these clients to new vistas and possibilities—a vision of a life of liberation and meaning. Generally, the father played an unusual role in inspiring the child, manifesting behavior that was irresponsible by conventional standards. One of the challenges for these types is to be able to live a fulfilling and meaningful life in society. They may well choose to live on the fringe because of frustration and dissatisfaction with conventional life styles. They need to cultivate a spiritual element in their life, or at least a social idealism, otherwise they are afflicted by meaninglessness and the sense of being a misunderstood outsider, which engenders corresponding delusions of martyrdom. At best, they can be a figurehead for the spirit of the times, and a source of great inspiration to others. At worst, they can seduce others into a delusional world that puts them at odds with society. On an existential level, they have an appreciation that reality is what you make it, which means that if their sense of self is strong, they can have tremendous self-awareness, whereas if it is weak, they can easily engage in self-delusion. Almost all combinations between personal planets and Uranus-Neptune give spiritual powers that can be developed into clairvoyance.

Moon to Uranus-Neptune

Those born with this combination lead unconventional lives in unusual environments. Generally, they experience the family as very unstructured, often with a mother who goes her own way, completely flouting maternal conventions. They have the perception that the mother refused to be stereotyped by marriage and may indeed have gone off on her own quest instead of baking cookies at home. With this aspect combination, there is often a sense of equality between child and mother, which the mother actively promotes, but at the cost of the security and dependence that most children need. This ambivalence is reflected in the later lifestyle, in which the Moon-Uranus-Neptune person rejects conventional domestic scenarios and embraces some kind of gypsy existence. They may live in foreign countries, and often there are periods in their life when they experiment with the idea of communal living. These people function better in group relationships than in one-to-one relationships. They feel uncomfortable bonding with another on an intimate level, and this leaves partners disoriented, confused, and sad. These people need to be made aware of their distancing techniques, which spring from lack of cohesiveness in early family life.

Mercury to Uranus-Neptune

This combination brings equal doses of genius and delusion. The person's mind is supersensitive and incredibly intuitive, with a special talent for clairvoyance, astrology, and many alternative subjects. They have a great hunger for inspiration and a corresponding boredom with traditional learning. This can well mean that early schooling was problematic, unless the person's artistic talents were encouraged. It is difficult for others to follow the thinking and reasoning of these types, and they often deliberately create a confusing mental atmosphere to avoid being pinned down into specifying their ideas. They are not renowned for practical thinking, and they are strongly reluctant to be brought down to earth. Often they lose direction and can spend periods of their life in a very confused state that is seriously exacerbated by drugs or alcohol. Some form of meditation practice can help them regain the clarity they need to get their viewpoint across. Once they have received spiritual guidance, they have the capacity to guide others through realms of consciousness.

Venus to Uranus-Neptune

For those who have Venus configured with Uranus-Neptune, relationships are characterized by experimentation that tends to lead to impermanence and disappointment. The person's early childhood experiences connected to sisters and the parental relationship instilled in them a lack of faith in the possible benefits of marriage or a stable love life. Boredom strikes quickly and is followed by restlessness composed of disappointment in the partner and the desire to try something new. These types may be advocates of free love and will see themselves as paving the way toward creating new and better forms of relating, unencumbered by tradition. Generally, they have at least one significant relationship with an exotic foreigner, involving much travel and separation, creating a mood of pining for the distant lover. The real challenges come when they enter a formal relationship and routine sets in. Ultimately, they find happiness through involvement with group ventures, especially if there is a spiritual component. They have an abundance of love and compassion that needs an outlet beyond the personal to bring pure fulfillment. No one partner can satisfy this need.

Mars to Uranus-Neptune

Clients with this planetary combination generally have very special talents that come to fruition through unusual careers that can range from electronic ability through social activism and into spiritual practice. However, they tend to be afflicted by sudden changes of direction, which are often the result of a profound dissatisfaction with long-term career demands. Restriction and monotony is unbearable for them. They have a pioneering spirit that often leads to long-distance travel and an attraction to exotic environments. Men with this aspect experience great challenges in coming to terms with their sexual needs, which are rarely conventional and, when indulged, rarely fully satisfying. They have a longing for transcendence through sexuality, which in turn leads to experimentation. Dissatisfaction with these experiments can lead to a conscious choice to follow a spiritual route. Women tend to be attracted to unusual and unconventional men who do not care to make a commitment, or cannot, because they live at a great distance or because there is a cultural divide. Happiness for both sexes comes when they work toward an unselfish vision they can believe in.

Uranus-Pluto: Ideological Revolution

While there are sides to this combination that exhibit a genius quite out of the ordinary, difficult combinations between these two planets create terrifying extremes in society. As Uranus approaches Pluto, ideological fanaticism grips the collective mind, the young guard overthrows the old order, and ordinary people, caught in the enormous upheaval that follows, suffer. So it was during the Reign of Terror in 18th-century revolutionary France (the conjunction), in Nazi Germany in the early 1930s (the square), and in the Cultural Revolution of China in the mid 1960s (the conjunction). At the same time, there is often a technological revolution, which is equally violent in its consequences, though quite necessary for the evolution of society. Those people born at these epochal times—and many people are—will channel the extreme revolutionary impulses to a greater or lesser agree. When they have personal planets making difficult aspects to the conjunction, opposition, semi/sesquisquare, or square of Uranus and Pluto, they probably have undergone a serious shock or traumatic event that has rent the psyche. This shock needs healing, and the first step to healing is for clients to share.

Sun to Uranus-Pluto

People with this combination will tend to experience considerable alienation from society, probably living on the fringe in some way. Often they will embrace and live out some future archetype that shakes up complacency in others, provoking censure and judgment. Their urge to provoke is strong and bordering on the self-destructive. The relationship with the father is likely to be difficult. Sudden separation and unpredictable—even suddenly violent—behavior perceived in others will underline the client's impression that anything can happen, any time, and it could turn nasty. Their reaction to adversity is to precipitate drastic change with no thought for the consequences. Transcendent experiences elevate their consciousness to the plane of the gods. Spiritual dangers threaten their sanity, but ultimately, dramatic consciousness change empowers the individual.

Moon to Uranus-Pluto

Upheavals in childhood rip away cozy family foundations and leave Moon-Uranus-Pluto types with the sensation of the ground suddenly opening beneath their feet. The mother in no way conforms to the usual maternal archetypes. She may come from another culture and have an aura of alienation. More often, the child experiences the mother as a person with an extremely fragile psychological state. The child senses the yawning chasm of psychic instability and flees from the abyss. Subsequent anxiety impedes settling down into a stable family life. The threat of moving hangs in the air. Partners are driven desperate by the person's habit of freaking out and indulging in emotional striptease, and in extreme cases, others will feel they are walking on egg shells. It's essential for the person to learn emotional self-control and to deal with psychological issues with a therapist or in group work.

Mercury to Uranus-Pluto

Though there may be considerable genius with this combination, especially in advanced technological, digital, and scientific areas, these types have disruptive communication patterns that create unease in the people they are with. They may have endured extreme and shocking events connected with siblings, and also as they went through the education system. Sometimes a deadly secret that they dare not reveal—perhaps of a sexual nature—puts the fear of God in them. In their youth, they may have engaged in some form of dysfunctional behavior, and the cure imposed by the parents or society may well have been worse than the problem. They have acute anxiety about their mental state, a sense that the border between sanity and madness is wafer thin. Sometimes they learn to play on this with a kind of zany humor. Communication with them tends to be driven and intense.

Venus to Uranus-Pluto

People with this combination have very unusual relationships throughout their life. They are scarred by the sudden loss of love as

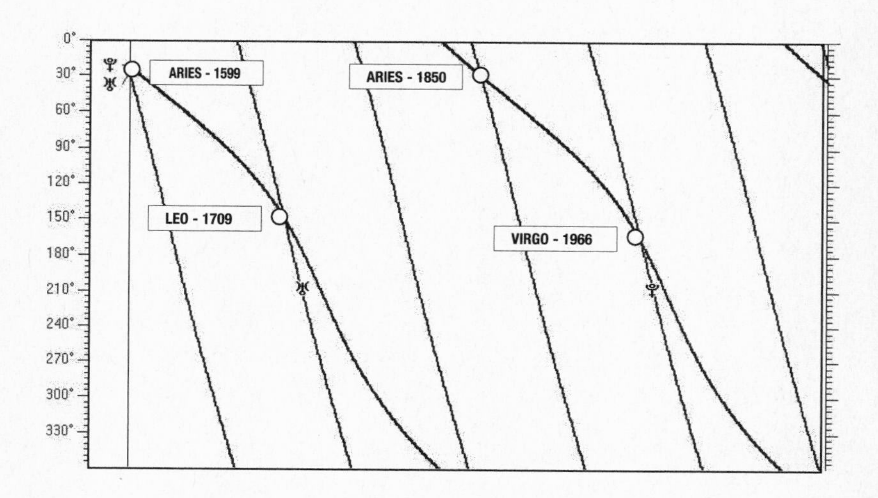

Fig. 28. Uranus-Pluto conjunctions 1600–2000. This cycle varies from 100 to 140 years. It brings dramatic, often traumatic, change, with fanatical processes violently replacing the old with the new.

children—connected either to the parent's relationship or to events regarding sisters or half-sisters—and subsequently anesthetize themselves. There is a strong likelihood of sexual experimentation, with intimacy suddenly established and just as suddenly torn apart. Both sexes send the signal that they have no real intention of bonding. Men with this combination evoke extreme insecurity in the partner, who can start to develop paranoia or self-destructive traits through sheer desperation. Women tend to exacerbate a deep conviction of worthlessness through the inability to commit for any length of time. They have a sense of extreme dissociation from the body.

Mars to Uranus-Pluto

While these people have a great talent for working with pioneering developments, especially of a technological character, their capacity for working in harmony with others is limited. They are outsiders and need to work to their own rhythms. People with this combination experience terrifying extremes, often connected with the presence of acute danger or ever-present threats as a child, particularly from an unpredictable brother or half-brother or the father or stepfather. Both sexes tend to develop extremely provocative behavior and are fearless to the point of self-destruction. Women are attracted to foreigners or outsiders, and permanent instability and extreme psychological pressure often characterize the relationship. Men have difficulty dealing with their sexuality, which is an unpredictable driving force that leads them astray.

Neptune-Pluto: Transformation of the Spiritual Ideal

This planetary cycle, the slowest-developing of them all, relates to the transformative power of a new vision of reality that grips the psyche of civilization and propels it onto a new level. The last conjunction took place around eight degrees in Gemini in 1892, at about the time of revolutions in communication—technologically via motorization, telegraph, and film—and on an inner level through dreams and psychology. Since the late 1940s and well into

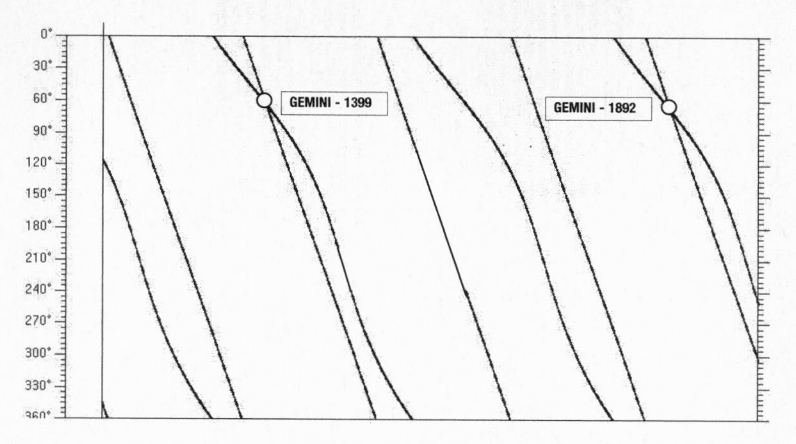

Fig. 29. Neptune-Pluto conjunctions 1200–2000. These conjunctions take place every 492 years and are associated with long cycles of spiritual renewal, connected with new visions of iconic power. For several millennia, they have taken place in Gemini.

---

this century, Neptune forms the outgoing sextile aspect to Pluto. Because of Pluto's wide elliptical orbit, (it actually came within the orbit of Neptune in the 1980s and 1990s), the first Neptune-Pluto sextile is well within orb for 100 years, and almost everyone alive will have this aspect in their chart.

As it is a positive aspect, it does not have a violently negative effect, the most difficult combination probably being the double quincunx finger of God aspect, or yod. Applying close squares to Pluto will also be quincunx Neptune, and separating close squares to Neptune will be quincunx Pluto. As this applies in all cases, it's difficult for the astrologer to discriminate between them, for their energies merge. The possible negative effect of the conjunction of a personal planet with either Neptune or Pluto is somewhat mitigated by the sextile, although beneficial aspects to outer planets are mixed blessings at best. The combination confers great imaginative power, and sometimes iconic status—these people channel a universal energy that strikes a chord of unconscious recognition in other people. Pluto and Neptune give radically different messages, as Neptune wishes to merge, accept, empathize, and surrender, while Pluto fights for survival when under threat.

Sun to Neptune-Pluto

While the person with this aspect has the potential for accessing healing power and for tremendous spiritual transformation, in times of crisis they have a tenuous hold on identity. Their nature is split by self-obsession on the one hand and self-sacrifice on the other. Somehow the father makes a weak impression as an individual—though he may be quite autocratic—and he is universalized into something more than he is or replaced by an idea or vision. The person's urge for spiritual insight can be strong and all-encompassing, and often the greatest discoveries are made at times of greatest destruction. These people will follow a path that few others can, and they defy conventional wisdom.

Moon to Neptune-Pluto

When these individuals have sufficient resources, they are compassionate and effective caregivers. More often, though, they are fighting their way out of the slough of despondency or through an emotional crisis of one sort or another. They seek refuge in dreams and fantasies, but ultimately any illusions they have are stripped away. A difficult relationship to a mother whom they may have perceived as a martyr or somewhat disturbed leads to an affinity for complex relationships that tend to be too intense, or just unhappy. This person has considerable instinctive and intuitive ability that gives them psychological and healing talent.

Mercury to Neptune-Pluto

People with this planetary combination will have unusual mental powers and can become skilled investigators both of the inner world of the psyche as well as the outer world of persuasive power. They are inclined to be secretive and careful or selective with information, sometimes building up a view of the world that is suddenly shown to be empty or false. They have skills that are appreciated in the advertising industry. With an instinctive knowledge that things can become what you say they are—in the sense that mud sticks—they can be led astray and lose their grip on what is real. To attain happiness they should avoid mental pollution from without (too much input confuses) and self-delusion from within.

Venus to Neptune-Pluto

People with this combination tend to have fateful relationships that range from unrequited love to intense and addictive dramas. Early perception of the parental relationship and possibly jealousies or losses related to sisters and half-sisters can lead to a conviction that love that is true will be lost, while the love that lasts may be unsatisfactory. Living out this inner script, women can cast themselves into inappropriate relationships, which lead to humiliation, while men tend to be dissatisfied with what they have and let themselves be drawn into secret fantasies about taboo liaisons. Both sexes are drawn to relationships that are bigger than they can handle, perhaps because of the need for transformation and sublimation through love and sex.

Mars to Neptune-Pluto

While this combination can give tremendous talent in fields that range from healing to politics, these people run the risk of undoing their efforts by engaging in self-destructive behavior and by their inability to rein in desire. Women with this combination tend to long for a special man, masterful yet sensitive, but end up disappointed and disillusioned by men who turn out to be wimps. The men have great difficulty finding channels for a sex drive that tends to be more in control of them than vice versa. They often have a sense of having a peculiar mission or of being driven by a vision. The more these people channel this energy into the collective—for example, as social concern—and the less they live it out in physical drives, the more satisfied they will be.

PART THREE

Transformation Methods

Introduction

Astrology That Works

While, for the sake of convenience, I will present in this section a range of therapeutic options in separate chapters, you will find that different strategies will work with different people at different times. As a general rule, if you're not making progress with one line of investigation, it's a good idea to drop it and try a new one. What is important is that you maintain rapport with the client, making interventions sensitively and cautiously. Unless the client is enrolled into the idea of making change, none of the strategies here will be very effective. If, on the other hand, you have clearly explained to the client the astrology of behavior and consequences, then your client will more likely be eager to embark on the adventure of change. Consultations done in this way will be exhilarating, both for the client and you, and will be a constant source of learning. The best teachers astrologers can have are their clients.

I would like to reiterate that the techniques in the following chapters are intended for basically healthy individuals on a journey of personal growth and insight—a journey that you as an astrologer can facilitate greatly. You will not cure serious trauma or personality flaws with these astrological techniques—far better to leave these aspects to those who have the specific training—but you will be able to make a dramatic difference in the lives of those seekers who end up choosing you as their astrologer.

You may feel that some of the work described in this section requires psychological training, and there is no doubt that it is a great advantage. There are many courses that can teach you communication and intervention techniques, and I personally have benefited from both my interest in NLP and Buddhism. However, psychology and astrology are two different subjects, and you have to choose which you wish to excel in. Astrology requires a slightly different set of skills and talents, and it takes time and practice to become good at it. Apart from knowledge and experience of the subject, it is most crucial to have interest in and care for the client. Studies show that the biggest transformative factor in therapy is the therapist, not the kind of therapy performed. If your motivation is pure, then you will have good results.

It is not difficult to have a pure motivation if you understand that the horoscope represents your client's contract with life. Taking on immense burdens and challenges, each client lives an extraordinary life, uniquely determined by his or her karma. It's a privilege that they have sought you out and that you are in a position to make a difference. If your motivation is pure, then you will not make mistakes. You may lack both astrological and psychological experience, but if your agenda is the client's good, then you will have a positive effect.

Many people with exceptional astrological skills find themselves working in other fields, either because of the financial considerations or because they feel they are not good enough. But after the requisite years of study, the only way forward is to practice with clients. Skills improve immeasurably when working with real people with real needs. When you take the plunge, you will often find that astrology can be a job like any other and can be the source of reasonable income. It may have been virtually impossible to make a living from astrology 50 years ago, but in the 21st century, the role of the astrologer is growing. If astrology is what you want to do, you should go for it—the universe will create space for you.

10 First Impressions— Winning the Client's Confidence

In the introductory phase of the consultation, it is important to win the confidence of the client, who often does not know what to expect from an astrologer and probably has little idea how far-reaching the subject of astrology is and how revealing the horoscope can be. Resisting the urge to launch into an interpretation of the most glaring difficult aspect, you are advised to establish your credentials by identifying, say, three issues that are important for the client there and then. To do this, you can use transits, progressions, and, if you have the skill, the consultation chart. The issues that are current for the client are almost invariably paralleled in the transits at the exact time and place of the consultation, so you'll often find, for example, a person with a Mars-Saturn contact coming when Mars and Saturn are actually aspecting each other. This indicates that there is a link between what is *currently* going on and what has *always* been going on in the person's life. By choosing the aspects that are echoed at the moment of the consultation, you greatly enhance the possibility of effecting change during the consultation.

Using the Consultation Chart

The consultation chart is a study in itself and warrants a complete book that details its use. It is a horary chart for the arrival of the client, and, indeed, this has been a centuries-long tradition. The great horary practitioner William Lilly, for example, was just as likely to only use the horary chart for the client's arrival as he was the birth chart. I have heard tell of astrologers in India who use nothing more than a stick. When the client arrives they implant their staff in the ground and, observing the shadow cast by the sun, know the exact time of the day. Being completely cognizant of the current position of all the planets, they know which sign is rising, and they are therefore already completely familiar with the consultation chart. From the shadow of their staff, they can relate details of the past, present, and future of their visitor. Astrologers today—being one-eyed prophets in a kingdom of the blind—use their computers as a crutch to perform a similar task.

Actually, it is quite possible to perform the same skills as these astrologers from India—and I know a number of astrologers who do—because working daily with clients gives one familiarity with planetary positions and the Ascendant sign at any given time of the day. Practitioners can perform a consultation without knowing any details of the birth time, date, and place of the person who crosses their path. In this scenario, the astrologer is like a wandering question mark—a wandering Descendant in which people entering the scene are represented by the Ascendant and its ruler. Every moment can be likened to a horary question, and in it is contained the past, present, and the seeds of the future. This practice makes every moment interesting, every person and event fascinating, and it enables the practitioner to use daily life as the most profound teacher of astrology.

Some psychologically oriented astrologers are disinclined to use the consultation chart because of its associations with traditional horary, which is often equated with a fated approach and to questions that require a yes or no answer. When working with psychological issues, it is unhelpful to suggest that things are fated. This is not the kind of horary I am advocating here; rather I am talking about a kind of "astrology of the moment." Not to realize that the consultation chart encapsulates the essence of the client's situation betrays a failure to understand the very nature of time. There is actually no way that this chart could *not* reflect the client's situation.

In traditional horary, some practitioners maintain that there are circumstances in which a chart cannot be interpreted. The word "strictures" has been coined to signify astrological configurations that prevent interpretation. These do not apply in the consultation chart. If the client turns up with the Ascendant in the last degrees or the first degrees of a sign, for example, it merely reflects something important about their situation (that is, whether something is being wound up or a new situation is about to develop). The same applies to whether the Moon is void of course, whether Saturn is in the 1st or 7th house, and all the other considerations that might traditionally be thought to prevent judgment. There are no circumstances in which the consultation chart for the client's arrival can *not* be interpreted. Otherwise, interpretation rules are similar to traditional horary: the client is represented by the Ascendant ruler, his partner by the Descendant ruler, his boss by the 10th-house ruler, his home by the 4th-house ruler, and so on.

The time you choose for this chart is the moment of meeting, of eye contact, of handshaking, whether clients come early, late, or on time. If a couple turn up for the consultation, then in all probability the one who appears to take the strongest initiative will be represented by the Ascendant ruler, and the more passive one by the Descendant ruler. For astrologers who practice on the phone (and here the consultation chart is invaluable), it's the moment there is two-way communication. If you choose to start giving advice to a client on the phone, and then later make an appointment, then both charts may have relevance, but the face-to-face meeting chart will take precedence. As telephone consultations are a growing field for astrologers, with new commercial services springing up all the time both on premium telephone lines and the

Internet, the crucial question as to what coordinates should be used—the astrologer's or the client's—poses itself. Centuries ago, this problem hardly arose, although traditionally, the client's coordinates were used for when a letter was received. Personally, being resident in Britain with most of my clients in Denmark, I experimented for six months using both sets of coordinates. There is now no doubt in my mind that the client's coordinates work best and that in the event of telephone or Internet consultations the horoscope should be relocated.

Some astrologers might object that the consultation chart is yet another chart—along with transits, progressions, and solar returns—that only serve to complicate the consultation process. Of course, this chart *is* merely a very detailed transit chart that has meaning in and of itself. It is much easier to use this along with the natal chart than to use progressions or the solar return, for example, simply because it gives a stunningly accurate depiction of the client's current situation. If, for example, the client has transiting Neptune square his Sun in the birth chart, Neptune's aspects and house position in the consultation chart will show exactly how and where this influence is manifested. And, as mentioned earlier, if aspect patterns are repeated in this chart—and some invariably are—then it is 100 percent certain that this reflects issues the client is currently struggling with.

With a knowledge, then, of the client's natal chart, along with transits and progressions, and hopefully armed with the skills to interpret the chart for the moment, it is good to start the consultation process by identifying key issues as precisely as possible. A large proportion of clients don't *really* believe astrology is truly capable of revealing their inner and outer life. It is a complete surprise for them when they find it can. However, if you, as the astrologer, fail to impress at an early stage, it can take a long time to win the client's confidence, and without confidence it is difficult to proceed with the psychological techniques that are described later on in this section. Conversely, if you reveal impressive skills of interpretation right at the start of the consultation, the client is very receptive to psychological change later on. As many of the techniques in this section are *questioning* techniques, it helps that the client believes you already have the answer, and to a certain extent, when you ask a question, you should have a good idea of what the answer will be.

Empowerment and Insight

Many people have been struggling for years with seemingly irresolvable problems and may well have visited therapists and other health experts in an effort to overcome them. The danger is that this work empowers the problems rather than the solutions. It would be a shame for the astrologer to further entrench these problems by giving them an astrological identity, cementing them for all time as a cosmic signature of suffering. If you start to work on a very difficult aspect—for example a Moon-Saturn conjunction in Scorpio in the 8th house—without having built up a positively charged atmosphere, the client will not have the resources to handle what comes up. As a matter of self-preservation, the client will not allow him- or herself to really access what is going on, and only surface issues will be covered.

Therefore, it is helpful to build up the power and confidence of the client by evoking the energy states connected with positive aspects and strongly placed planets. These influences have far more strength than most astrological textbooks attribute to them. They represent states of being that are extremely positively charged. Successful work evoking these states will empower the client, and even you, the astrologer, will enjoy them. As explained earlier in chapter 3, it is quite straightforward for the astrologer to elicit empowerment and happiness in the client via the positive influences in the chart. This must be the starting point for dealing with negative issues.

Of course, it can be a struggle just to get the client to see or admit to a problem, and if no problem is identified, no solution is possible. After you have done work to help empower the client, then it is crucial to present the behavioral patterns that cause him or her trouble. All so-called negative aspects represent a difficult integration of energy, and this energy manifests as an attitude, and this attitude creates a behavior, and this behavior has consequences—and these consequences are problematical. It is utopian to imagine that aspects such as squares and oppositions do not give specific problems. Generally, conjunctions involving Mars, Saturn, Neptune, Uranus, and Pluto, and all squares and oppositions (as well as semi- and sesquisquares) will give rise to specific behavior and cause easily identifiable problems.

Having identified a difficult aspect pattern, the next stage is to describe its manifestation in a way that the client can identify with. The best way to do this is to choose a difficult aspect that is currently shown in the transits for the day (and therefore in the consultation chart) and that is also present in the natal chart.

In figure 31 (p. 223), two visits from the client, with a year and a month in between, replicated her natal Sun-Jupiter conjunction straddling the 9th house (see figure 30, p. 222). Each time, the consultation chart picked up the theme of seeking happiness with lovers abroad, probably indicated by Jupiter ruling the natal 5th house. It related to the fact that her father used to take her on holidays—traditionally, they visited a popular island resort (Pisces)—as "a special treat," but in reality, it was so the father could get away from the mother and conduct a secret affair with another woman in what was a yearly ritual. The daughter, with Mercury in Pisces in the 8th, opposite Pluto-Uranus, kept this deadly secret throughout her childhood, but not without exacting material quid pro quo from her father over the years. Obviously deeply impressed by this recipe for happiness, the daughter had been replicating it ever since. On each of her visits to me she was involved with a married man in another country, who she hoped would leave his wife for her. After all, with Saturn on her natal 7th-house cusp, where was the appeal in being married?

It's quite extraordinary to see each feature of the Sun-Jupiter influence reflected in the story. The Sun in the 8th house, yet conjoining the 9th, showed the taboo sexual nature of the foreign attraction. And, apart from the island paradise theme, Pisces showed the unattainable dream and unrequited love, which provoked an existential crisis every time, and not just for her. Actually, it was with some alarm that I noted the repetition of the Sun-Jupiter conjunction for the second consultation. Surely if I had dealt with the issue successfully the first time, I would never have been presented with this repeat placement? It served to emphasize, at least, how crucially important this conjunction was.

The advantage of this method of working on current patterns is that you can be very specific about what is going on right now in the client's life. What is happening is that clients have a certain behavior that is reflected by a natal aspect. This behavior will influence their environment throughout their whole life. The current transits merely show the consequences of that behavior right now. By identifying these consequences it is much simpler to correlate them with earlier behavior and childhood character and events.

Astrologers are often tempted to describe childhood influences and events based on the birth chart, and clients are only too willing to tell long stories about their childhood. But unless a correlation is established between childhood and adulthood, all this information leads nowhere. When a correlation is established, it dawns on clients that they have had lifelong behavioral patterns that are directly responsible for problems they are currently experiencing. This is a very powerful realization.

By describing the influence of an aspect, then showing how this related to childhood events and behavior developed as a consequence of these events, and then correlating this behavior with current problems, the urge for change arises within the client. Unless the client expresses an urge to change things, very little can be done. Ideally, your description should be so effective that the client spontaneously bursts out with the question: "What can I do to change this?" It could be the woman with Mars-Neptune who repeatedly finds herself being betrayed by and philandering man; a man with Venus-Neptune who secretly nurtures the dream of the

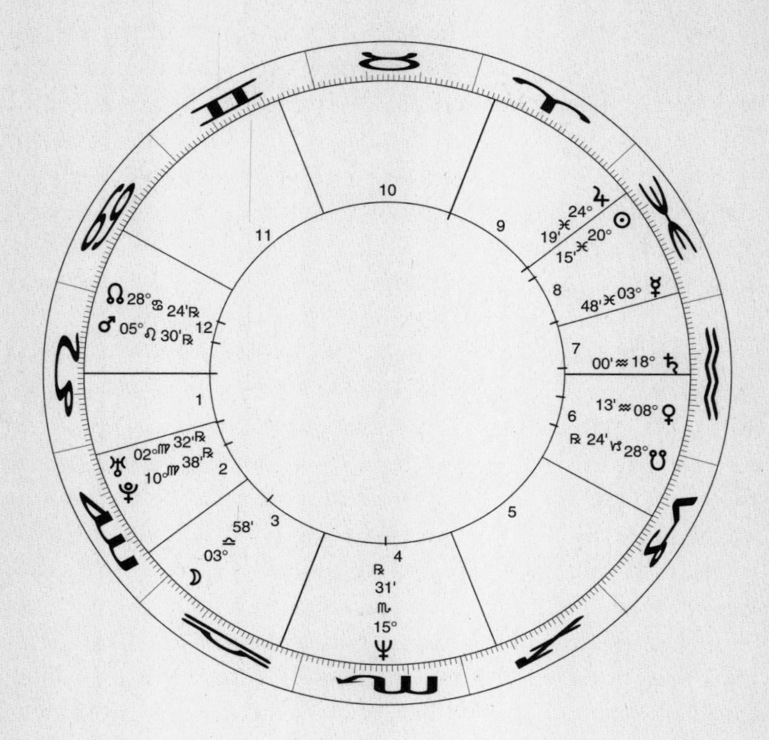

Fig. 30. Charlotte D. Note the parallels between this birth chart and the charts for the consultations shown in figure 31.

Fig. 31 (p. 223). Two consultations. An interval of one year shows almost exactly the same Sun-Jupiter conjunction position, straddling the 9th-house cusp.

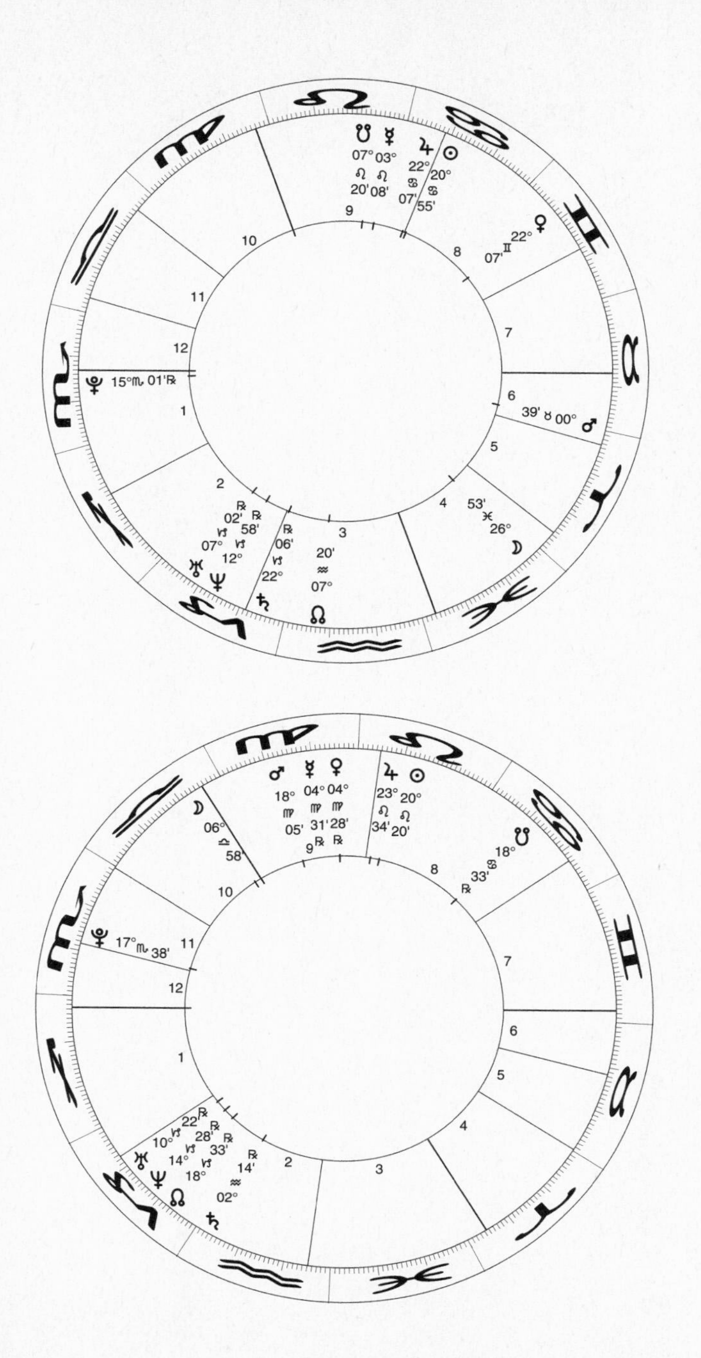
☋ ☿
07° 03°
♌ ♌
20' 08'
♃ ☉
22° 20°
♋ ♋
07' 55'
22° ♀
07' ♊
♇ 15°♏ 01'℞
39' ♉ 00° ♂
53'
♓
26°
☽
℞
02' ℞
♑ 58'
07° ♑
12°
♅ ♆
℞
06'
♑
22°
♄
20'
♒
07°
☊
1
2
3
4
5
6
7
8
9
10
11
12
♂ ☿ ♀
18° 04° 04°
♍ ♍ ♍
05' 31' 28'
℞ ℞
♃ ☉
23° 20°
♌ ♌
34' 20'
☽
06°
♎
58'
☋
18°
♋
33'
℞
♇ 17°♏ 38'
22℞
10° ♑ ℞
28' ℞
♑ 33'
14° ♑ ℞
18° 14'
♒
02°
♅ ♆ ☊
♄

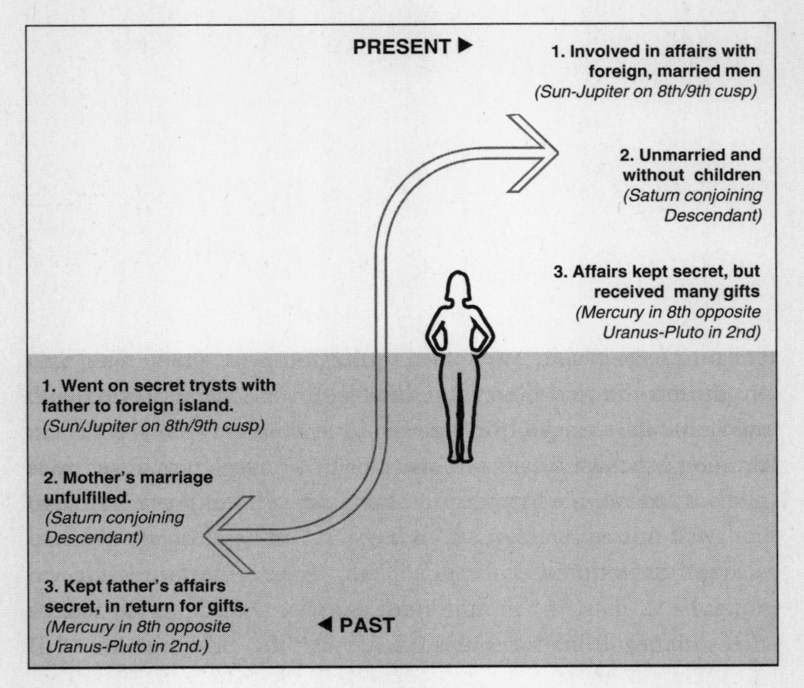

Fig. 32. Parallels between past and present. Core events in the past will cast long echoes into the future, often spawning unconscious actions that eerily replicate the original event. The transit chart for the consultation will help pick out these repetitions.

---

perfect relationship, while his partner languishes and fades into neurotic unhappiness; or an individual with Mercury-Pluto who drives everyone into silence with insistent and overly intense communication. If you can successfully identify how this manifests now and what the causes from childhood were and establish the correlation, then the client *will* ask the crucial question, "What can I do to change this?" Awareness of the problem is the first step to transformation.

Awareness of the Problem—and Seeing It in a New Way 11

Everyone has difficult aspects, and the behavior associated with them is almost always embarrassing. People would prefer to keep it secret. Yet a precondition to change is owning and taking responsibility for this behavior. Later, in chapter 15, I shall show how the guilt connected with this behavior can be alleviated and even transformed. In the meantime, the challenge is to get to the point where a person acknowledges a behavioral pattern and asks for help to make changes. To get to that point is the first triumph. Getting there is a combination of knowing your astrology and having effective communication techniques. These techniques, described earlier, help the astrologer get to the core issues in a very short space of time. With this combination of astrology and communication, it can take from one to five minutes to get to a core issue, though it can take considerably longer to evoke it so strongly that the client asks for help to change.

Expanding Awareness of the "Problem"

Many clients would feel more comfortable if they could gloss over their problems—wouldn't we all? One way they do this is they quickly admit to the problem in an off-hand way, thereby hoping to dismiss it. Others project the problem onto others, and, indeed, when planets are in the 6th, 7th, and 8th houses or when oppositions are involved, it seems often to be the case that other people do apparently manifest the problem, rather than the client. And there are some clients who totally deny that there is a problem. As an astrologer, you know that if the difficult aspect is there, then there is trouble somewhere. Sometimes you do meet people who have dealt with and transformed problems so that they no longer manifest. It is rare, but it happens and, indeed, it is the aim of the astrologer to expedite the client into this last category of people. Transformation is possible, and some people attain it. When transformation has happened, then there is no need any more to use the word "problem," and in fact this word should be used sparingly, if at all, in the consultation. I use it in this book for convenience only.

When you have successfully described the manifestation of a difficult aspect, then the client will concur to a greater or lesser extent. To get a sense of the ramifications of the problem, it is possible to evoke a much stronger sensation of it using techniques such as those described in chapter 3, "Inner Resources." This is done by building up a stronger sensory experience. When clients map out the sensory signals connected with a problem, they come to a basic energy that connects them with all the manifestations of the problem through life, often bringing them back to a core experience early in childhood. Remember, this is a very powerful technique, so you should use it *sparingly*, and only when the client does not really seem to connect with the negative behavior.

An example of this could be a Moon-Saturn conjunction in Scorpio square Uranus. This could be identified as a situation in which a woman client experiences a desperate alienation in her current relationship, which could be correlated to a time when her mother went through a period of mental instability in which she was unable to relate to the client as a child. The consequence might therefore be that the child amputated herself emotionally from the mother in order to cope. Now if the client was only to partially relate to this scenario, then sensations could be strengthened by evoking sensory states connected with it. Thus:

Astrologer: And when your mother ignored you, how long could that be for?

Client: A long time.

Astrologer: How long specifically?

Client: Maybe two weeks or so.

Astrologer: And when this happened, what did you feel?

Client: Nothing at all. [Amputation]

Astrologer: And when you felt nothing, did you have any particular sensations?

Client: Well, I felt cold, and alone.

Astrologer: Was that a particular sensation in your body, then?

Client: In my tummy.

Astrologer: Like a vacuum, or some kind of knotted feeling, or what?

Client: It's like an emptiness.

And you could go on, with each question strengthening the sensation. With such a technique however, there is a danger of inducing the very trauma the client has been trying to handle every since. While this is quite harmless with negative aspects to Jupiter, or softer aspects generally, with planets like Saturn and particularly Pluto it is dangerous to go on too long. When you deem that the client is strongly enough in contact with the consequences, then you have to extricate them, which you might do in this kind of way:

Astrologer: And is this the same sort of emptiness you feel when you are alienated in your current relationship?

Client: Yes!

Astrologer: So there's a connection, right?

Client: Yes!

Astrologer: So isn't it possible that the alienation you feel in your current relationship has just as much to do with your behavior as your partner's?

It is at this point that you could expect the client to understand that the responsibility for a healthy relationship was hers and that she could do something about it, so that she might ask, "Yes. What can I do about it?"

With a diagnostic tool as efficient as astrology, it's only a matter of time before you have mapped out various problems shown by the chart. Typically, the client will be quite familiar with these problems. They will have tried various methods for solving them, yet they will go on manifesting. Now that you have successfully identified the problem in so short a space of time, the client will look at you expectantly. Perhaps you have the solution?

It's easier said than done, and for a number of very good reasons. Often a difficult planetary aspect pattern will represent an extremely complex set of behaviors that form the backbone of a person's life experience. At the core of this is a basic energy that is immensely powerful. The difficulty of making change should not be underestimated. Effecting change under these circumstances can be likened to shifting the flow of a mountain torrent that for years has carved its way through the landscape, creating its own valleys and landmarks. It's difficult to do, but it's not impossible. However, it is much easier to do if you go to the source: giving advice of alternative behaviors is going to have little effect, while rechanneling the energy at its very core has a greater chance of success. The more entrenched the problem, the more you have to go to the core energy.

A New Look at Old Problems

People are very attached to their so-called problems. Often they will have developed creative strategies for dealing with them that become crucial parts of their career. A man with Sun square Mars-Saturn may find that the resilience he built up to handle a violent father is just the quality he needed to have success in the corridors of power later in life. Someone with Mercury square Neptune may find that the smokescreens used to overcome learning difficulties at school are just the thing for being a top salesman today. These difficult aspects are transformed into winning formulas that bring success. However, when people build success on the indiscriminate use of such strategies, they often end up with a sense of dissatisfaction or obsession about the work involved.

Respect for the Problem

On an even deeper level, people see their problems as a crucial part of their identity. I recall an episode I experienced in India when I was 21. My first child had just been born in Goa, and at the age of one month, he was unfortunately unable to be breast-fed any longer and had to be bottle-fed, which necessitated the purchase of a paraffin stove to heat the milk. This stove was cursed. It had three legs and a plunger that had to be vigorously pumped to generate the flame. Alas, one of the legs was weak and would periodically fall off, and this meant wandering around various Indian cities in our travels looking for a place to get the leg welded back on. I was unerringly directed by helpful Indians in exactly the opposite direction of the welding shop, and with early summer temperatures of 95° Fahrenheit and above, fixing my two-legged stove rapidly became a nightmare. On these occasions I would curse the world, India, and myself in particular, for ever buying the stove.

In Kathmandu, the leg fell off again, after holding up for at least a month and lulling me into a false sense of security. I set off to find a welder, and as usual, I was misdirected around the city and ended up taking refuge in the dingy recesses of a local café, bemoaning my fate. While sitting there, an Indian holy man—a *sadhu*—made his way through the café to the table where I was seated nursing my anger. He pointed to the stove, tilted precariously on the tabletop on its two good legs. He asked me if I would like to sell the stove. I stared at him. He offered me 30 rupees for it, just a few rupees less than I had paid. Would I sell it? He picked up the stove, repeating his offer. No way! No way was I going to sell that stove to him after carting it all over India and lovingly getting its bad leg welded on time and again. I took the stove back. That was *my* problem, and no one was going to take it away just like that! It has amused me to think that the stove, which I energetically pumped to provide heat to make food for my family, described the basic dilemma of a Cancer with Moon in Aries, as it was the cause of so many bad-tempered episodes.

Considering this refusal to divest myself of an item that for four months was the bane of my life, I have developed another kind of respect for the difficulties that recur in a person's life. Eastern philosophies resolutely refuse to see problems as problems, but instead view them as a treasure chest containing costly items on which the individual can practice again and again, until the lessons have been learned and the karma exhausted. When dealing with clients, this approach pays dividends, because it takes the sting out of problems and removes the embarrassment connected with the associated behavior.

Self-Acceptance

The first stage of transforming problems into resources is self-acceptance. Given the fact that everyone, without exception, struggles periodically with and often conceals acutely embarrassing behavior, there is nothing special about the client in this department. He or she is just like everybody else. It's a funny thing, though, because the very secret that the client strives to conceal is often of only passing interest to anyone else. People who admit to their problems are in a much stronger position than people who don't. So the first point is that there is a distinct advantage in sharing a problem, because sharing purges it of its power. Secrets wield such power, but when you, as the astrologer, unearth these secrets and show that they share the generic qualities of many other people's secrets, you help the client defuse this power.

One client boasted to me that he had seduced over a hundred women. Man to man, he expressed some pride in this. Yet, he was afflicted with a debilitating jealousy in case his wife should be unfaithful while he was out on his travels. This was probably connected with the Moon-Pluto conjunction in Leo in the 6th—seduction at work was obviously part of the job—and the stellium of planets in Libra, combined with a weak Mars conjoining the 5th-house cusp (see figure 33, p. 232).

He made himself jealous in the following manner: he would imagine his wife meeting a man at a hotel bar, drinking with him, and ending up in his hotel room. (No points for guessing where he got this little scenario!) He had an extremely vivid imagination, and he would imagine the whole episode running like a full-color movie from start to finish. He did this compulsively, and he could not stop doing it. It had reached the point where it had almost taken over his life, and he would hardly let his wife out of the house. I dealt effectively with this in the consultation by asking him to run through the film he created faster and faster, and then run it in reverse, which is an NLP technique for dealing with phobias.[9] However, the point here was that he was so ashamed that he was in the thrall of his jealousy and had no control over his imagination. Yet this is generic for jealousy. Almost everyone suffers from jealousy to a greater or lesser extent, and they often create their suffering by directing their own vivid inner movies, imagining a fantasy scenario. Having done this, they then despise themselves for their weak behavior.

Someone who is afflicted with a bad temper will often use an enormous amount of effort trying not to lose their temper, only to suddenly explode. Then they get furious at themselves for getting

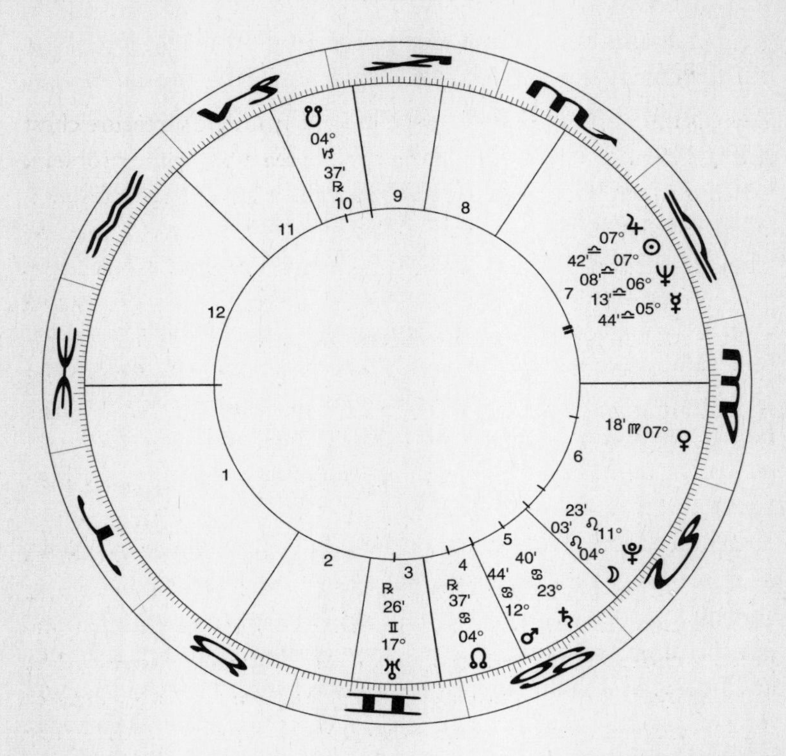

Fig. 33. Stan. The Moon-Pluto conjunction in Leo and the 6th, combined with the womanizer tendencies of the Libra stellium, made this man a "master of seduction," but left him victim to jealous fantasies.

---

angry. Someone who lacks intellectual self-confidence may spend hours trying to master a subject, only to be shown up as ignorant when the crunch comes. They then indulge in an inner critical dialogue in which they demean their own intellectual capacity. What happens with every class of problem is that it *happens*, then people hammer themselves for letting it happen. Jealous people despise their weakness, angry people are furious at themselves, weak people further undermine themselves, greedy people indulge in self-hate, and in doing all this they compound the problem.

Self-Respect

There is the original problem, which comes from life's treasure chest and is difficult to resolve; then there is the reaction to the problem, which is much, much easier to resolve. This is a good place for you, as the astrologer, to start. The problem is revealed in the horoscope through the different planetary configurations, which, at their core, show a powerful energy that drives specific behaviors. The reaction to the problem is added later and comes because people feel there is something wrong with them. Given the fact that everyone has something similar, the idea that something is wrong is false. It is the human condition.

Furthermore, this person is *born* with the planetary configuration. It could be argued that this person has in fact chosen this specific energy to deal with in life. The Venus-Pluto man has chosen to be born with a compulsive need to seduce, which he will transform. The Mercury-Saturn individual, terrified of being thought of as stupid, has chosen to work with ignorance and attain betterment. The more extreme the manifestation of the problem, the more respect the individual deserves. A person with Venus square Neptune may have a sickly sister, may be deserted in love, may lose money through naiveté, and may run through the whole spectrum of this aspect's manifestations during his or her life. To choose to work with such difficult issues is fantastic! Having chosen a life that in part will be governed by a difficult planetary configuration, many events will be spawned. Once you fully understand this, it is easy to feel enormous respect for your client.

This respect is very empowering for clients. It enables them to cut out the 50 percent of the suffering that otherwise would have come from the secondary reaction to inappropriate behavior. A person is jealous: so what? Many people are. Angry—that's right, but at least working on it. Ignorant, yes, but getting more intelligent, day by day. It's the astrologer's acceptance and respect that is important here. The person has a problem; it's something he chose to deal with in this life—that's fantastic; he's getting more and more expert at dealing with it as he develops, and one day, one day soon, it will be cracked.

Self-respect could be evoked in many different ways; the important thing is that you have an appreciation for what the client has gone through, and you show it.

Astrologer: So you've been with over a hundred women, but you have had an acute problem with jealousy?

Client: It would seem so.

Astrologer: And there's a connection between these two things, right?

Client: I guess so.

Astrologer: And would it be true to say that even from early childhood you'd thought obsessively about seduction, even up to this moment? That it had kind of taken over. [The use of verb tenses is meant to root this behavior in the past, rather than as something ongoing.]

Client : Yes.

Astrologer: Do you see this as something that you want to continue throughout your life?

Client: No, I do not. [Emphatically]

Astrologer: One thing I've noticed is that you're very conscious of this behavior and that you're actively working on it, especially in this consultation, aren't you?

Client: I would like to overcome it.

Astrologer: I'm sure there are a lot of people out there that have the same kind of problems but are not at all as open as you are. Also, wouldn't you agree that a person who went through the same conditions in childhood would have to confront the same challenges? [He had a philandering father and a mother who told him of her own affairs.]

Client: It's possible, I guess.

Astrologer: So it's quite understandable for things to develop this way for you. You're working on it—indeed you're something of an expert on this kind of problem—and you expect to crack it, right?

Client: Right!

The therapeutic element of this exchange is that the astrologer shows a nonjudgmental interest and respect and injects a sense of evolution into a state, which, until then, seemed permanent and deteriorating. It's not going to make a big difference in his behavior yet, but the next time he seduces a woman, he'll start making connections and will begin to develop alternative strategies. And the next time he's afflicted by all-consuming jealousy, he is less likely to beat himself up about it.

Reframing—an Empowering Intervention 12

Reframing is a therapeutic strategy that is very mild—it can never harm—and you can use it in the consultation without any specific psychological training. It is based on the idea that if you present an event or a behavior in another light, it suddenly transforms from something negative to something positive. It is easy to see when reframing has succeeded and when it has not. If it hasn't worked, the client responds intellectually and undramatically; if it has worked the client is surprised, exhilarated, and empowered, and this can be seen in the whole body language. Figure 34 on page 239 shows some reframing arguments that correspond to problems associated with each of the ten planets.

In life, things happen, and then people make up stories that make what happens conform to their preferred way of seeing the world. It's the same old story. Your girlfriend leaves and goes to live with your best friend, and you, with Mars in the 11th house, square Neptune, learn once again that you can't trust friends. An employee is fired, and, with a Saturn-Pluto conjunction in the 10th, he knows that bosses are just out to get him. The parents divorce, the father apparently makes little effort to be with the Sun-Pluto daughter, and she grows up to despise the weakness of men. These are subjective interpretations of events that simply happen. There are many alternative interpretations, but none of them occurs to the person involved.

It's actually a very interesting exercise, in a class of students, to elicit alternative explanations. Taking the Sun-Pluto daughter as an example, the students proposed that the father visited once a month because:

- The child's mother actively sabotaged visits.
- He thought it best not to upset the child with too many visits.
- He did not want to disturb the equilibrium of his ex-wife's new marriage.
- The child herself had said she was bored when with the father.
- His new wife got mad when the child came around.
- He simply did not like children and was happy to have his freedom.
- All her friends lived near her mother, so she actually did not want to visit him.

Listening to a host of suggestions from other people, the person in question will often look incredulous. Are there really so many alternative viewpoints? He or she only has space in the imagination for one.

This awareness—that no interpretation has the patent on being correct—forms the basis of the successful reframe. An event happens, and the best interpretation of it is the one that most empowers the client. The only criterion for a reframe's relevance is whether it hits home. You may think that you've suggested an excellent alternative

Planet	Problems associated with planet	Arguments enabling a reframe
☉	**Identity Problems**	Only through crisis is it possible to get to know oneself on a deep level. Depth and resilience.
☽	**Anxiety and insecurity**	Through overcoming insecurity, it is possible to find an inner home that nobody can take away. The ability to empathize is developed.
☿	**Contact problems**	Developing a serious attitude to communication means depth instead of superficiality. Knowledge hard-won has depth and durability.
♀	**Low self-worth**	When self-worth is built up through own efforts, there is self-sufficiency. No dependence on others for self-confirmation. Values come from within, not from without.
♂	**Anger and fear**	It's better to feel something passionately than indifferently. It's more important to be honest and direct than to be liked. Winning can be crucial.
♃	**Limiting convictions**	Beliefs are crucial for the future of society. Having a strong belief is a sign of searching for the truth. Trying to convince others is connected with the urge for them to improve.
♄	**Repressions**	It is through difficulty that one develops strength. When things do not come easily, they have greater value when attained. Self-discipline gives power and control.
♅	**Alienation**	Someone has to show an alternative way forward, and that can mean being an outsider. People will not change without a shock being administered.
♆	**Indefinable sorrow**	There are spiritual gains from being in deep contact with the feelings. It's commendabe to search for the ineffable. The desire for the ideal is first step on spiritual road.
♇	**Power/Powerlessness**	Great power comes from meeting inner demons. Having experienced great extremes, there is nothing to be afraid of, either in oneself or in others.

Fig. 34. Angles for reframing. Each planet presents its own opportunities for constructing a reframe. The result of any difficult aspect always contains positive possibilities, which can be identified and used in the reframe.

view, but if there is no reaction from the client, it should be dropped and a new one tried. The main question to ask yourself is this:

What would *not* have happened if this event had *not* taken place?

or

What has this event resulted in that the client would not be without today?

Any event has consequences of which the person involved strives to make the best, and generally there are always some good results.

One 58-year-old client, born to a rich family on a Caribbean island, complained bitterly to me of an episode that took place when she was 15. At this time—to the horror of her well-to-do parents—she fell in love with a man from a poor family and wanted to marry him. Frustrating relationships at the age of 15 are the hallmark of Venus-Saturn aspects, seen in figure 35 as the opposition between Virgo and Pisces on the 5th/11th-house axis.

Her parents acted promptly. They took her out of her local school, and sent her to a private college in the United States. They intercepted all communication between the love birds and effectively ended the relationship. She had never forgiven her parents for doing this, and had only on very rare occasions ever returned to see her family. In this, she made an error. Was she going to have a happy love life, given the Venus-Saturn configuration and the emphasis on Pisces on the one side and Neptune on the other? The unhappy affair at 15 was actually an accident waiting to happen. Her sorrow and regret, indulged for one and a half Saturn cycles, was simply a manifestation of the powerful energies in the configuration focused on the profound emotional loss she experienced as an adolescent, which was an echo of an older, deeper loss.

What actually happened was that she led a privileged life, traveling to America, joining the diplomatic corps, and getting posted around the world. The States, the Far East, London, Paris—I met

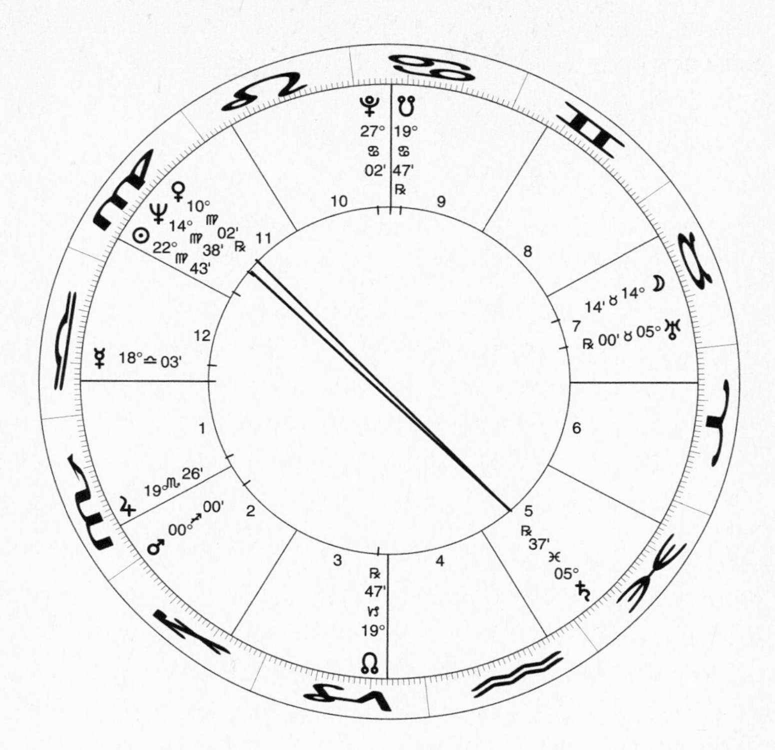

Fig. 35. Miranda. As Venus retrograded and stationed by progression opposite Saturn in puberty, a "tragic love affair" happened that changed her attitude toward her parents for many years.

---

her in Copenhagen, where she was currently stationed. Her exalted Moon in Taurus, opposing Jupiter and conjoining Uranus, showed that she gravitated to a life of upheaval and relocation and also of luxury, and this was not a pattern that would have acclimatized to a life locked away on a Caribbean island. The powerful 7th-house ruler, Mars, on the Scorpio-Sagittarius cusp, trining Pluto, also showed a preference for cultured and powerful men. It is hard to see her settling in with a man in a rather inferior position and with herself taking on a subservient role in the long term.

This, then, was the angle to take regarding the reframe—focusing on the positive side of living a privileged, international life, as opposed to a life as a slave to a limiting relationship.

Client: I can never forgive them [the parents] for that.

Astrologer: You wanted them to let you marry your boyfriend at that time?

Client: Yes.

Astrologer: What's he doing now; do you know?

Client: I heard from my sister that he's still living in the village, and he's married with a lot of children.

Astrologer: [Ironic] And you could have had all that!

Client: Well that's not the way it went.

Astrologer: No. You've traveled the world, speak several languages fluently, meet influential people—you've had quite an exciting life, haven't you?

Client: Yes, I really have.

Astrologer: So let me just check this with you. You'd rather have married at 16 and stayed in this village during your life, as opposed to educating yourself in the United States and traveling the world?

Client: [Thirty-second silence. Something powerful is happening inside.]

Astrologer: It seems to me your parents may have done the wrong thing, but it had the right results. No?

This interchange had an extremely profound effect on the client, who subsequently was able to return home and forgive her parents and reconcile herself to a life that she had, in reality, chosen and could not take responsibility for. She chose to give up seeing herself as a martyr for love, with her parents as the bogeymen. Indeed, she thanked me profusely for laying an old ghost to rest.

Another client I had nursed a bitter hatred for the police, reflected in a general way through his Mars-Pluto opposition from the 2nd to 8th house. It "sprang" from a time when he was imprisoned for dealing with hard drugs, when he was himself a user. It was difficult to see this strong man in front of me as an erstwhile addict, but so it was. It probably had a lot to do with identity problems connected with hard aspects to his Sun. The police used his addiction to extract information from him about dealers, actually allowing him to fix when he cooperated and withdrawing the drug when he did not. This was, of course, an illegal interrogation technique. He found the experience so utterly humiliating that he actually kicked the habit so that he could not be manipulated in this way. But he hated the police, and particular members of the force, with a vengeance. What was really happening, though, was that his potent Mars-Pluto opposition, which earlier was connected with the self-destructive process of heroin addiction, had now attached itself to the police. It would be healthier if he were more aware of this energy, and this inspired the following reframe:

Astrologer: So you were actually given heroin in jail to make you inform on your suppliers?

Client: Yes, and later there was a scandal about my case.

Astrologer: How long had you been an addict?

Client: Four years.

Astrologer: But now you teach skills to kids with behavioral problems, right?

Client: Yes, I've been doing that for a few years now.

Astrologer: So what were you doing at the time you were addicted to heroin?

Client: Well, nothing. Got money to buy my next fix. It was hell.

Astrologer: Why didn't you just kick the habit?

Client: [Pitying my ignorance] It's not that easy.

Astrologer: What was the name of the detective that interrogated you using this heroin bribe?

Client: A jerk called Roy X.

Astrologer: I'd like to meet this Roy and shake his hand.

Client: What!?

Astrologer: Yes, of course. He accomplished in two weeks what you had failed at for four years. If you hadn't been so furious at being manipulated [Mars-Pluto hates humiliation], you'd never have gotten your act together, right?

As mentioned before, you can always tell when a reframe hits the mark. The client goes through quite a powerful and often wordless process. There's little to say afterward. It's as if a row of dominoes falls down and a mirror of perception is held up that changes something forever. Bitterness or sorrow is often replaced by gratefulness, or at least appreciation.

Memory—Helping Clients Recall Core Material 13

Memory is mystery. In the storehouse of the mind, memories exist that traverse generations and dynasties. In the maelstrom of every moment, we selectively filter out extraneous information, concentrating on the information we need to get what we want at any given time. What we accept and what we reject is based on our mental dynamics as reflected by the horoscope. Much of what happens around us is scarcely committed to memory, simply because it does not resonate with our needs or our nature.

What we *do* remember resonates strongly with the dynamics of our being in some way—if it didn't, we wouldn't remember it. Memory is, therefore, far from so-called objective truth; it is more a definition of who we are as souls. There is an old Russian saying, "Never trust an eye witness"—this reflects a view of how selective memory actually is. Some totalitarian ideologies have deemed it necessary to reinvent the past to create a new identity for their nation, creating, as it were, new collective "memory." Others see that in unearthing the past and protecting tradition, they create a strong and authentic national identity by making memories part of the collective myth.

Some theories of reincarnation suggest that identity is a conglomerate of embedded and quite disparate energies that are reassembled into selfhood with the coming of consciousness, only to dissipate again upon unconsciousness, or death. In this sense, the memory of who one is at any given moment is the force that constantly recreates the sense of identity. At birth, the baby rests in this pool of memory—so much to get reacquainted with—and slowly gives definition to identity in the interaction with parents, siblings, and the world. It is at this point that residual memory affects behavior and begins to define the environment. Thus, the infant with a powerful Mars-Uranus triumphantly spits out the pacifier—placed in his mouth to quiet his riotous nature—inviting retribution from the harassed parental authority (the first stifling of the right to rebel). Residual memories of starvation and survival prompt the Moon-Pluto child to suckle as if his life depended on it, evoking stronger than usual reactions in the mother. (Or sometimes milk means poison or powerlessness.)

In this way, children gravitate to themes that will become dominant in their lives, and the ethereal, disparate energies that make up the karmic drive for selfhood consolidate into apparent identity. Personal memory is added to and embellished with a collection of episodes that reinforce existing patterns. Much successful therapy is based on the reinterpretation of remembered events in the light of a wider perspective that is not warped by an emotional and subjective viewpoint. With this view, no other person can be fully responsible for the difficulties experienced by the individual in life. Indeed others are "co-opted" into one's own existential drama in a complex interactivity whereby people get together to fulfill development needs and drift apart when they are fulfilled. Gratefulness, forgiveness, and appreciation replace blame or guilt when it is realized how others fulfill one's own needs in this way, and this can be very cathartic.

An important consultation skill is to reveal to clients how their inner dynamics evoke specific behavior in others. In any long-term relationship, we will be apparent victims of behavioral patterns in others that will frustrate or annoy us, but these are patterns that we are partly or totally responsible for. One client with Venus in Pisces in the 7th had two consecutive relationships during which his partners both became genuinely ill. So what did the client do to evoke this? On investigation it turned out that while he wasn't generally very romantic, when the partner was ill, he was extremely solicitous and was happy to sacrifice his time to devote himself lovingly to the partner. Obviously, both women felt that this was worth getting ill for. But it is not a very empowering love life, and the problem can be solved by consistently showing caring behavior, particularly when the partner is healthy. This would naturally require recognizing and transforming the guilt patterns, which were accessed through residual memories of suffering women.

Accessing Memory

When investigating early memories, it is common to hear clients claim that they do not recall events. Some will even say something like, "I don't remember anything before I was eight." This is never true. Either something was committed to memory because there was resonance with it, or it wasn't. Nothing of importance is actually forgotten. What happens is that people choose not to remember for one reason or another. When people say they *can't* remember, what they really mean is that they *won't* remember. There may be a good reason for this, and there are times when the client's "nonmemories" should be respected and left alone, but there are many occasions when it's crucial to recall the past.

It is astonishingly easy to access these nonmemories. Let's suppose you're working with a female client with a strong Moon-Saturn aspect. The first step—which works about 70 percent of the time—is as follows:

Client: I don't remember anything before I was eight.

Astrologer: What exactly don't you remember for example?

Client: I have no memory of my mother holding me or cuddling me.

Astrologer: So what did she do when you went to be cuddled.

Client: She rejected me.

Astrologer: How, specifically, did she reject you?

Client: She was always too busy cooking, looking after my father, going out to work . . .

So, basically, here you simply register that the client thinks she can't remember, ignore her claim, and ask what it is precisely she cannot remember. It could go like this:

Client: I don't remember anything before I was eight.

Astrologer: What exactly don't you remember, for example?

Client: As I said, I don't remember.

Astrologer: So, if you could remember something, what might it be?

Client: Well, I wasn't so happy.

Astrologer: In connection with your mother, your father, or what?

Client: My mother was never very warm.

And this will lead into the same scenario. This will work about 20 percent of the time when the first technique does not work. You simply ask, "If you could remember, what might it be?" This might not work either, and it could go like this:

Client: I don't remember anything before I was eight.

Astrologer: What exactly don't you remember, for example?

Client: As I said, I don't remember.

Astrologer: So if you could remember something, what might it be?

Client: I've told you—I don't remember anything!

Astrologer: OK. But imagine a girl about eight—like you—what sort of things would you expect that girl to remember?

Client: I simply don't know.

Astrologer: Just make something up. Imagine an eight-year-old girl and tell me what she might recall.

Client: Well, I guess she might go on vacation with her parents.

Astrologer: Did you do that?

Client: No, my friends did, but we didn't have the money.

And this will lead to the same scenario. It should be said that the more a person claims not to be able to remember, the more interesting it is and the more rewarding accessing these memories will be. It really helps when accessing memory to use your powers of suggestion. Unconscious processes in clients make them very responsive to direct commands. These commands can quite simply be embedded in your questions, for example:

- If it were true that *you can remember this episode* . . .
- If you were to *recall this episode* . . .

- Do you think that *you unconsciously have all the details of this in your memory*?

The italics show a suggestion intoned as a command.

Using these techniques, and combining them with embedded commands, you can have a certain knowledge that you *will* be able to access whatever memories are necessary for the work at hand. Clients will remember 95 percent of the stuff they think they have forgotten.

14 Past, Present, Future—Changing with the Time Continuum

It is a very common practice in the consultation for the astrologer to delve deeply into past issues, especially the client's relationship with the parents. However, in and of itself, finding out about the past serves no useful purpose. Going through childhood memories of a neglectful mother or an absent father merely serves to open up old wounds. What does work is to weave the significance of the past into the client's present situation and to plan a strategy. In dealing with every astrological issue, it pays to weave past, present, and future into a continuum.

Significant difficult aspects manifested strongly in childhood, they manifest strongly now, and they will manifest strongly in the future. In childhood, the client experienced the full force of the negativity in the form of difficult relationships and events; at the present time there are also identifiable difficulties and so, too, in the future. But people do improve; they do acquire skills that help them transform these difficulties. Sometimes these skills enshrine the negative energy as a powerful and effective tool, which brings good results yet perpetuates the problem. For example, a man with a Sun-Pluto conjunction in the 11th house might have success as a power broker, but he pursues his career filled with anxiety. No transformation takes place here. In other cases, people with the same pattern may work more consciously with the psychological issues and facilitate group insights via personal development courses. In these cases, an ongoing transformation is taking place. It makes sense to aim for optimum transformation and assume that the very fact that clients are seeking guidance places them in the category of people most likely to attain transformation.

The key factor related to past events is the consequences they have. This is the reason negative manifestations are perpetuated. The young child has an experience and then makes a decision or draws a conclusion based on the experience. What causes the trouble later on is the result of that decision or conclusion. The young 15-year-old girl with a Venus-Saturn square is disposed to experience rejection in her first love. At this early age, she probably has no awareness that it has anything to do with her own behavior and perceptions.

The ending of this early relationship may well be experienced as painful, and it is at this point that major decisions are made:

- "I won't declare my feelings for a man until I'm absolutely sure of his feelings for me."
- "I'll have a relationship with someone safe next time."
- "Relationships cause pain; I'm going to concentrate on professional success."

These strategies have consequences:

- The man she truly loves will move on because he receives no signals that his feelings are reciprocated.
- She marries a man unworthy of her and ends up feeling trapped or bored.
- She leads a successful but lonely life.

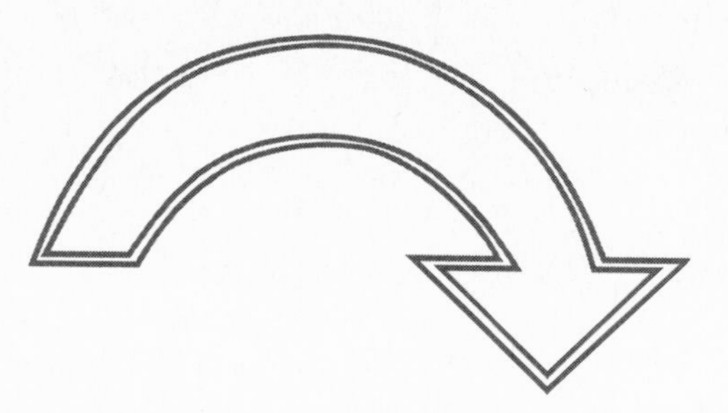

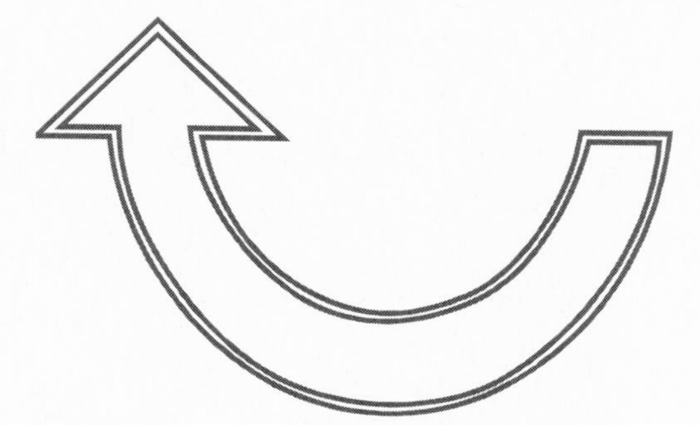

Fig. 36. Decisions and consequences. Using a Venus-Saturn aspect as an example, this diagram shows how the strategies used to avoid rejection will often serve to insure it. Early decisions need to be reevaluated in terms of their consequences, and changed.

In the consultation, the best introduction to a therapeutic intervention is to describe these consequences, because they are being manifested right now and because the client probably makes no connection between them and her behavior. Hearing a clear description of a current scenario is a riveting experience for clients, and they become very curious and more willing to open up emotionally. Once the consequences have been described, it's relatively easy to trace the feelings back to the early key experience. In the light of this, early experiences can be reevaluated. The client's decisions at that time may have worked, but they are demonstrably harmful now. Considering that the interpretation of what happened was strongly colored by selective perception dictated by the horoscope, then there is every reason to make a new and more empowering interpretation.

Having identified the current manifestation of a difficult planetary combination, traced it back to key events in the past, and ascertained what kind of limiting decisions or convictions were formed during that key period, you and your client can then look at future scenarios that will give greater fulfillment. The next step is to link up the new perceptions to a desired future scenario based on the simple question, "What do you want?"—a question that harnesses the visualizing power of Jupiter.

Creating the Future

It's an extraordinary thing, but the fewer resources people have, the more incapable they are of answering the simple question, "What do you want?" It seems that just to enunciate a future wish or desire requires energy, belief in oneself, and faith in the future. It demands an insuperable effort for those who have been bearing too heavy a burden for too long. When people do begin to describe what they actually want, their hearts immediately lift; they will raise their eyes, often looking up (to the right) and into the distance. What is happening here is that they actually start constructing a representation of the future, and usually it is visual. Clients have to mobilize the fire energy or Jupiterian energy in their horoscopes to entertain a prospect of the future. The very act of doing this is therapeutic.

Before this happens, though, you may have to overcome the client's resistance based on resignation and memories of disappointment. For example, a man with Sun in the 6th house in Capricorn, square Neptune in the 4th, may be depressed and disillusioned at work. When asked what he wants, he is only capable of seeing the reality of what he has got. Perhaps it has been going on for years and years—dragging himself off to work in the mornings, longing for weekends where he can get away. The "reality" is that he has to earn money, and his current job appears to be the only work he can get, even if it is undermining his emotional health. So it might go like this:

Present scenario: He is unhappy at work, as indicated by Sun square Neptune.

Past event: His father dreamed of running his own business but ended up working as a bank clerk. He drowned his sorrows in the pub and shirked family duties. The client had a deep sympathy for the father but saw him deteriorate as the years went by. The subliminal conviction: work makes you unhappy, but you've got to keep at it, even if you martyr yourself.

Astrologer: So how would you like it to be in the future if you could have it how you want?

Client: There's no point in thinking about it . . . I really don't know. There's just no other work around, and I have a family to support, anyway.

This type of response is quite normal, whether it's the wife who is stuck with a husband she does not love, a person who is resigned to never having a relationship, or someone who is stuck in a deadening job.

But the rewards of defining a positive future scenario are too great to let slip. If you can help your client remove certain negativities that stand in the way, you can stimulate the client to define a wish. In doing so, Jupiter gains ascendancy, while Saturn, which has been dominating the proceedings, takes a back seat. When people create a life for themselves, the building blocks are Jupiter and Saturn. As you wake in the morning, Jupiter clicks into gear and you automatically run through the things you want to do. It is Jupiter that gives hope and faith in the future. Jupiter represents the power of visualization: the enunciation of a wish and the sensory embellishment of that wish. People actually imagine future scenarios all the time, and these scenarios become very real. Jupiter *is* positive thinking. These constructs of the imagination are very powerful, and they magnetize the future, drawing you toward it. People with a strong Jupiter are motivated by their dreams and wishes and attain success through their magnetic power. People with a strong Saturn are more motivated by fear of failure than the allure of success.

When working with clients on future scenarios, you need to harness the visualization power of Jupiter, and the greater the sensory detail, the better. Clients need reassurance that it is possible to fulfill their wishes, and to do this, they need to see how their wishes correspond to their deepest personal values. Once the client has expressed a wish, then you can explore a future situation, making it as real as possible through the use of questions that engage as much of the client's sensory apparatus as possible. What would the client see/hear/feel/smell/touch in this future scenario? (What color? What's in the background? What noises? What else can you see/hear/feel/smell/touch?) In this way, the client creates a future that has the power to draw him or her into it. Basically, people *do* get what they want, simply because the power of wanting mobilizes Jupiter, and Jupiter is very, very strong.

There are two main reasons why people may not get what they want, and they both have to do with Saturn functions. One is simply a question of fate; there are times when you simply don't get what you want but get what you need. There is little that can be done about this. As the saying goes, "Help me to accept what I cannot change." The other is linked to sabotage mechanisms and nega-

VISION
- **Construction of future scenarios.**
- **Belief that they are possible.**
- **Imagination as to how things can be.**
- **The conviction that the desired future is in harmony with life values.**

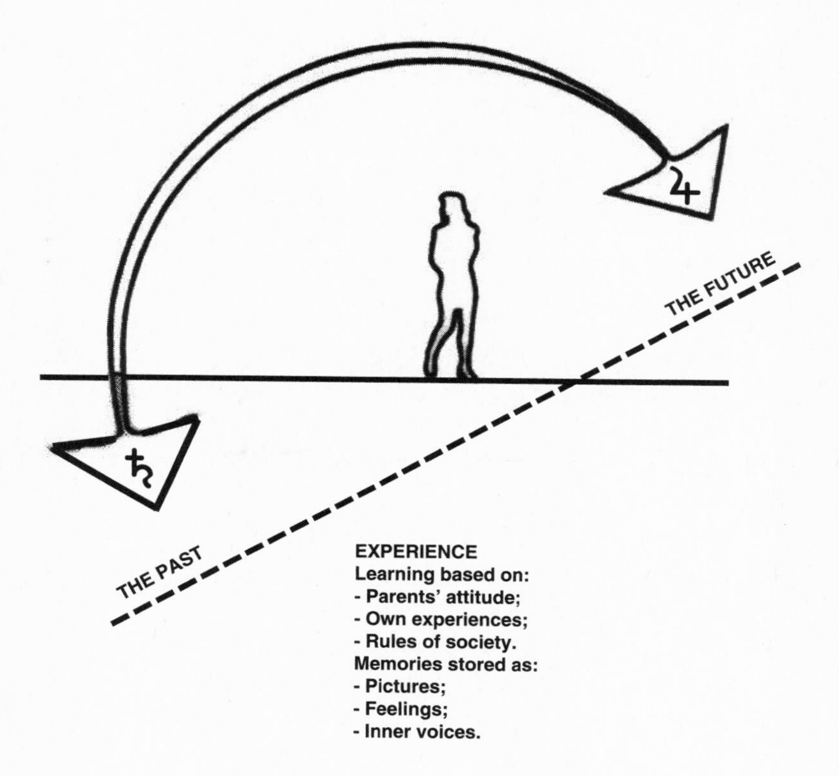

Fig. 37. Past and future: the interplay between Jupiter and Saturn. Experiences can limit vision. They are stored as inner pictures, feelings, and voices. When people construct future scenarios in their imagination, these old sensory states give rise to reservations about the possibility of achieving hopes and dreams. But if the vision accords with basic inner values, then it can be achieved. This is the interplay between Saturn and Jupiter.

tive thinking, and something *can* be done about this. When people express what they want, it will often happen that they feel some reservations. These should each be identified and dealt with in turn so that they do not undermine the power of the visualization. Reservations are based on past experiences, which are either personal or learned from parents or society. On an internal level, they are usually experienced visually as pictures, kinesthetically as feelings, or audibly as an inner voice.

Going back to the Sun square Neptune client, his objections to visualizing the future may very well stem from inherited attitudes typical of Capricorn.

Client: . . . and I have a family to support anyway.

Astrologer: Like your dad, then? Is that what he said?

Client: Yes, he did, actually.

Astrologer: And when *you hear his voice now*, can you hear the tone and expressions? [Embedding suggestion to hear father's voice. Whenever a client recalls something being said, the memory becomes very vivid when he or she connects in an audile manner.]

Client: His voice is frustrated and sad.

[Value judgments. It's more powerful to get a straight sensory description.]

Astrologer: Is the tone loud or soft, or what?

Client: It's subdued and tinged with emotion.

Astrologer: So that's the voice you use when you convince yourself you can't get what you want, is it?

Client: Yes. My dad's voice!

Astrologer: So what do *you* want?

For some people, goals are difficult to attain because they simply have not had any training in formulating and specifying goals. In these cases, their Jupiter may be strong but Saturn not well integrated. Learning to define and attain goals is not strictly an astrological skill—every self-respecting corporate self-development course works precisely on this issue, as well as on visualization and positive thinking. Corporate skills are very much based on Jupiter and Saturn archetypes. Nevertheless, it may be of use to your clients if you can help them organize their future. One of the biggest problems for some who strive to attain a goal—like, for example becoming an actor, or writing a book—is that it is not properly defined and is too vast. To attain any goal, smaller milestones have to be specified and attained. So, while it is good to vividly imagine the finished result because of its efficacy as a motivating factor, it also becomes necessary to organize milestones.

It's good to give a client homework in this respect, because formulating goals takes time. There are a number of golden rules about creating the future in terms of goals:

1. Goals need to be committed to paper, and they should be formulated in the present tense, as if they have already been attained.

2. They have to be measurable. In other words, it is not sufficient to say, "I am prosperous and happy." The client should write down the exact salary that he or she will earn. "Happy" must be measurable in concrete terms.

3. Goals must be specific, simple, and mean something for the client personally. It is not acceptable to say, "I live up to my wife's expectations."

4. Goals should be realistic, responsible, and ecological.

5. They should have a timetable and an end date.

6. They should be formulated positively. In other words, it is not acceptable to say, "I have stopped feeling bad about my work. Instead, the client should say, "I feel good about my work."

An example of these principles being fulfilled could be (for our Sun-Neptune client): "By January 2003, I am a freelance landscape gardener, earning $100,000 per year. I have two well-paid and loyal employees. I am living happily with my wife in a new five-bedroom house in my favorite Atlanta suburb." All these statements are measurable, specific, and have a timetable.

Authentic goals spring from the core of a person's identity—different people have very different goals. If sabotage mechanisms are no longer dominant, then there is every likelihood that people will attain their goals. It is their destiny and birthright.

Engaging the Client in the Prediction Process

Generally speaking, when helping clients define goals, you will be able to identify what is appropriate and possible. Transits and progressions will give a powerful indication of what is realistic in the coming years. For example, with Pluto approaching the Midheaven, it could be more relevant to encourage a client to totally abandon a dying career, while, with Saturn approaching, defining new responsibilities could be the way forward. The whole issue of the future is very delicate, indeed, because of the paradoxical situation you, as the astrologer, are in. It is effective to encourage the concept of free will because it motivates the client. Yet experience shows that it is possible to make an educated guess at what the future will bring. The last thing you want to do is to limit clients' options by describing a seemingly immutable future, yet, at the same time, it would betray the astrological tradition to foreswear prediction.

There is an ethical minefield here, but astrology is not served by avoiding prediction altogether. In India, there is a rule about prediction: "Don't predict anything bad for which you cannot suggest an antidote," which is quite useful. Indeed, old traditions use talismans, mantras, and many different rituals to allay the forces that might bring misfortune. I am reminded of one story in which a young woman who was about to wed visited an Indian astrologer who saw to his horror that her first husband would die only weeks after the marriage. The wily astrologer suggested he perform a ceremony in which she was married to a tree. Two months later, the tree withered and died, and the woman then safely married her husband—not the kind of strategy that would go down well in the West, perhaps.

The key thing to remember about the future is that the client already senses what it will be! Clients are perfectly capable of working out what is viable and what is not, and they will shrug off inaccurate predictions and embrace accurate ones. This is basically because of the function of Jupiter—the client is already actively (if unconsciously) visualizing the future, and when you describe a possible future scenario, it will either resonate or it won't. When it does resonate, this is the time to expound upon all the possibilities that the client senses and, in the last analysis, wants.

Another point to remember is that no matter how good an astrologer you may be, you cannot be sure how planetary energies will manifest and be handled by a client. For example, one married client had a Sun-Moon conjunction at 29° Libra in the 7th house, and when I asked what happened when the Saturn-Pluto conjunction transited this point in 1983, he replied that his hand was partially severed by a power saw. He spent a year out of work, and his wife suddenly had to go out to work. Their relationship was turned upside down, but it was apparently greatly strengthened too, not least by an episode that took place while they were on holiday in Thailand. They were in a small boat that capsized, and as the hours went by without rescue, they said goodbye to each other, only to be saved at the last moment by a fishing boat. That's quite a way to celebrate a Pluto transit! Personally, I would have thought no marriage could survive that transit unscathed, but quite the opposite happened. It is a constant surprise to see all the different manifestations of any particular transit. Therefore, as a rule, it is wise to engage the client in the prediction process and to use what *they* want and expect as a guideline. It makes sense to imagine what is the optimal manifestation of any transit and aim for that.

15 Deeper Intervention Techniques

Many of the interventions I have described so far are gentle intrusions, which can help clients on the road to self-discovery and change. During the course of a consultation, this kind of work may be all that is appropriate, and, indeed, all there is time for. However, the very nature of most difficult aspects is that they are a core energy at the center of being. It is utopian to imagine that they can be completely transformed through rational processes. Energy can only be transformed by energy.

Preparing the Safety Net

To access this energy, you need to be familiar with the four elements and their manifestation, described in chapter 3. When you know in which sensory system a behavioral problem is layered, then you can employ techniques that will enable you to pinpoint the problem quickly and immediately access the energy. However, I cannot stress strongly enough how potentially powerful this process is. You do not want to unleash a demon that you cannot control.

To guard against this happening, there are a few very effective techniques for extricating clients from the maelstrom of destructive emotion. Trauma can only be experienced when the client is "associated." This means that the client is directly experiencing, with one or more sensory systems, an inner representation of events. She feels herself to be in her body, looking out of her own eyes, and having strong body sensations. However, when a person is "dissociated," there is not the same vivid sensory experience. People who are dissociated see themselves experiencing events from the outside, as observers. Generally, the further away they are, the more dissociated. This is very useful to know, because traumatic memories are often best approached from a dissociated point of view until the client is comfortable enough to associate.

If, when dealing with trauma, clients suddenly begin to panic, it is because they begin to powerfully reexperience the event through their senses. You can calmly guide them out of this state in the following way:

Client: (Pale, breathing faster.): Oh no, this is terrible!

Astrologer: OK. Just take a look at yourself in this situation. Imagine seeing yourself as another person at some distance. Just tell me what she is doing now.

Client: He is holding me, I'm trying to . . .

Astrologer: [Interrupting] OK. He is holding *her*, and *she* is trying to what? Refer to yourself as "she," please.

Client: She is trying to pull her arms away.

Astrologer: So when *you are looking at yourself now* in this scenario, *how far are you away from the scene*? [Suggestion reinforcing dissociation]

Client: About two or three yards . . .

Astrologer: OK. Look at the scene from about five yards away.

This will effectively remove panic. You can go further by working on each of the sense systems if you need to. Generally—but not always—a scene in color and three dimensions is going to be more powerful in its effect than a black and white two-dimensional representation, so the client can be instructed to change the representation if necessary. It is truly amazing how events can be changed by altering the way they are represented via the five senses—sounds can be quieted, touch anesthetized, and smells transformed.

Sometimes you can set the scene up beforehand if you know you are going to get into traumatic territory and you want to keep things under control. One technique is to have the client imagine they are in their imagination on a cinema seat, and then simply ask them to play the scene on stage, in black and white if necessary. You can remove them even further to the projector room, where they have complete control of what they choose to show. This method is particularly useful when you are working with phobias. It enables people to run through the whole process without getting emotionally overwhelmed, giving them a chance to identify key causes of the phobia.

Some therapeutic training on your part is a precondition to working on the deep-seated energy behind traumatic experiences. Much can be achieved by simple astrological methods without this kind of deep work. However, it does happen that a long-term client has the courage to really get to the bottom of a difficult issue and expresses the desire to work in depth. If you both have the courage, and you have the training and experience, then you can proceed.

The Power of the Four Elements

In the naming of the four elements—earth, water, fire, and air—there is the tendency to make the assumption that these words actually refer to these elements as we know them in the outside world: the water of lakes and rain, the fire of flames, the earth as soil, and air as wind. In fact, these naturally occurring manifestations are all mixtures and, as such, low-level energy. Soil contains earth (its matter), water (its moisture), fire (its warmth), and air in the space between the particles. All these natural manifestations contain each other in different proportions.

What, then, are the elements? They are primeval forces that are all nonmaterial in nature, around which manifestation takes place. Earth is related to the impetus toward structure and solidity, and in the individual, contact with this energy gives a sense of peace and a feeling of being grounded, of being really comfortable in the body. Water is related to all fluidity and moisture, and contact with this energy gives a sense of being connected both within and outside of the body. Fire is related to the generation of warmth and is connected to temperature in the body and a primeval drive to want or go for something. Air is connected to space in the body on a cellular level and to a primeval drive for continuity. It is quite possible to experience these energies with meditation practice, though they have to be accessed experientially and not with the rational mind (see Appendix 1).

Accessing Elemental Energy

Deep-seated problems cannot be transformed through "normal," rational processes in the consultation. They spring from difficult planetary combinations, which have a history of events and experiences to reinforce them. These aspects are part of the energy and identity of the individual, who will not give them up lightly (just as I would not sell my paraffin stove in India). Nevertheless, the very energy that informs these planetary structures is transformational. The difficulty is that we experience transformation as the death of a part of Self, and that can be frightening. Only in dreams does this transformational energy get a chance to show its face, generally at times when Death is encountered in one or other of its forms. Generally, we flee from Death in all its guises while asleep, though certain lucid dreaming practices and dream yogas help practitioners develop the skills to work consciously with such fears in their dreams.

Every astrological combination has a root energy, which is activated whenever the consciousness of a client is directed toward its manifestation. The first step, then, is to accurately describe the manifestation. Central to this process is delivering an interpretation while keeping in mind which sensory system is most strongly stimulated. The rules are not hard and fast here, but you can expect problems centered in fire signs to be represented strongly through images and film sequences, in air signs through sounds and voices. Problems centered in earth signs manifest through physical sensation, and in water signs through deeper feelings rooted in the head and body. It's rarely as simple as that, especially because most difficulties are indicated by squares and oppositions, and therefore different sensory systems. When working with positive resources it's another matter, because they generally manifest through harmonious elements.

Expanding the Sensory Signals

Behavioral problems, then, manifest through specific sensory systems. Take for example squares between planets in Virgo and planets in Gemini (refer to figure 38, p. 269). Here it is a virtual certainty that the trauma is represented aurally through sounds and voices, simply because both signs are ruled by Mercury. Knowing a problem is represented verbally helps you get to the root energy very quickly:

Astrologer: It appears that you experienced your mother as very critical, is that right?

Client: Critical is not the word.

Astrologer: More than critical, then?

Client: She shouted at me all the time. She was mad.

Astrologer: So, when you *close your eyes* it's easy for you to *hear her voice*?

Client: Oh God! [Groans]

Astrologer: So is her voice loud, grating . . . what?

[Note: Value judgments (angry, sarcastic) are not important, we need to have an exact sensory description. The more the client simply concentrates on the quality of the sound, the deeper she will associate.]

Client: Yes, it's very loud, kind of staccato, the words are coming very fast. I'm frightened. She seems mad.

Coming to the core energy can be that fast. Here, the astrologer resists rationalizing: "You see, your critical mother is shown by this Moon square Uranus and Pluto." This would immediately dissociate the client from her emotions, (although this can be quite useful if that's what you want). Instead, the astrologer uses words and suggestive phrases that propel the client into a strong associative state, in this case concentrating on inner sound. There are several directions to go from here. One is to actually work with the sound, using the mind to transform the sound quality and induce another feeling.

Astrologer: What direction is this sound coming from . . . is it in your head or does it seem to be coming from the outside?

Client: It's right inside my head.

Astrologer: Well it's not your voice. Put it outside where it belongs. How does that feel?

Client: Much better. I feel more in control.

Astrologer: Right! How far does the voice have to go so that *it is so faint you can hardly hear it*? [Suggestion]

Client: Uh . . . that's amazing. Wow! That feels better!

It is surprisingly easy to move sounds, images, and feelings around in the inner and outer space which all people use to store them.

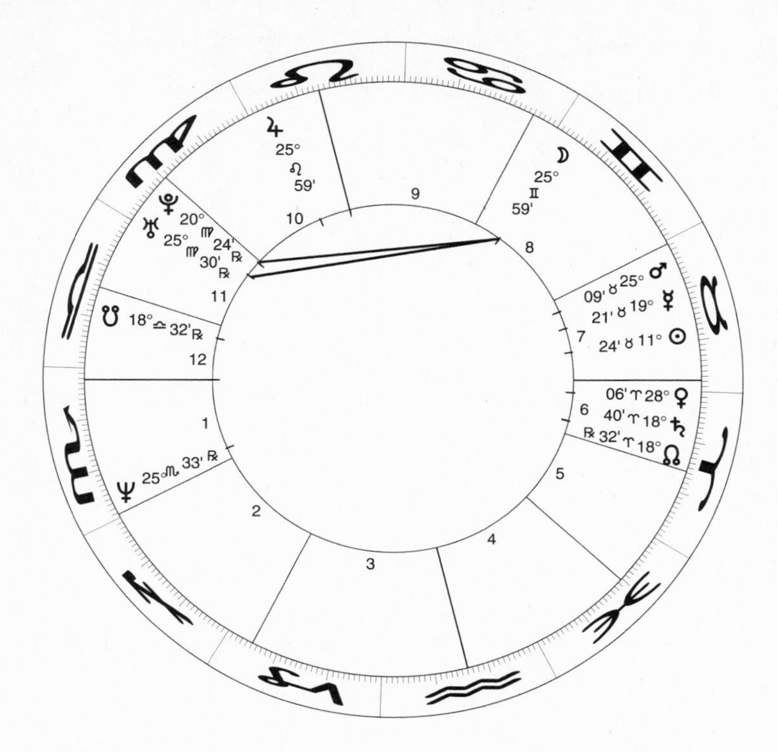

Fig. 38. Marianne G. The emphasis on Mercury, through the squares from Virgo to Gemini, coupled with the Moon's 8th-house position in an air sign, indicates that trauma is stored as an inner voice.

Each thing has its own exact coordinates, and these points of personal space, extending many yards around them (as well as internally in the body and head), are each associated with a particular feeling. Feelings change dramatically when representations are moved. For example, it is often the case that something placed behind you situates it out of sight and in the past, and generally, the closer a representation is, the stronger it will be felt. I remember once working with a person who had Venus retrograde in Scorpio conjoining Saturn—she felt an unbreakable and unhappy bond of love to someone in the past. She could summon a picture of him about four yards directly in front of her. She was able, under some protest, to experiment with moving him behind her and out of sight. When I saw her six months later I asked how he was. She looked at me as if I were nuts. "Oh, I put him behind me ages ago," she said, with an unconscious flip of her hand over her left shoulder. It's powerful and transformative to move sensory representations in this manner, so it needs to be done sensitively.

Though it is not always the case, people see the future ahead of them and the past behind them. A large number of people tend to see time represented as a line stretching from behind to the left, to in front and to the right.[10] Often this line is also inclined, ascending as it moves into the future. People with a strong Saturn will often see this line as very steep, while those with a strong Jupiter can gaze far along it. I've noticed many astrologers (and probably others, too) can see the past stretching in front of them to the left and the future in front of them to the right. Perhaps this is because they operate so much in past, present, and future simultaneously, they want to keep an eye on them all.

Taking a new example of accessing and expanding sensory signals, this time in another sensory system, a man with Mars in Aries square Saturn in Capricorn would be less likely to store his difficult memories in auditory mode. Beginning with the cue from where the *personal* planet is placed—in a Fire sign—it is more likely that there is a strong visual element to his memory function, probably combined with a body sensation connected with Saturn in earthy Capricorn. The exchange could go like this:

Astrologer: It seems like there was a heavy disciplinary element in your childhood . . . ?

Client: Well, my father was rather violent.

Astrologer: How often did he hit you?

Client: Regularly.

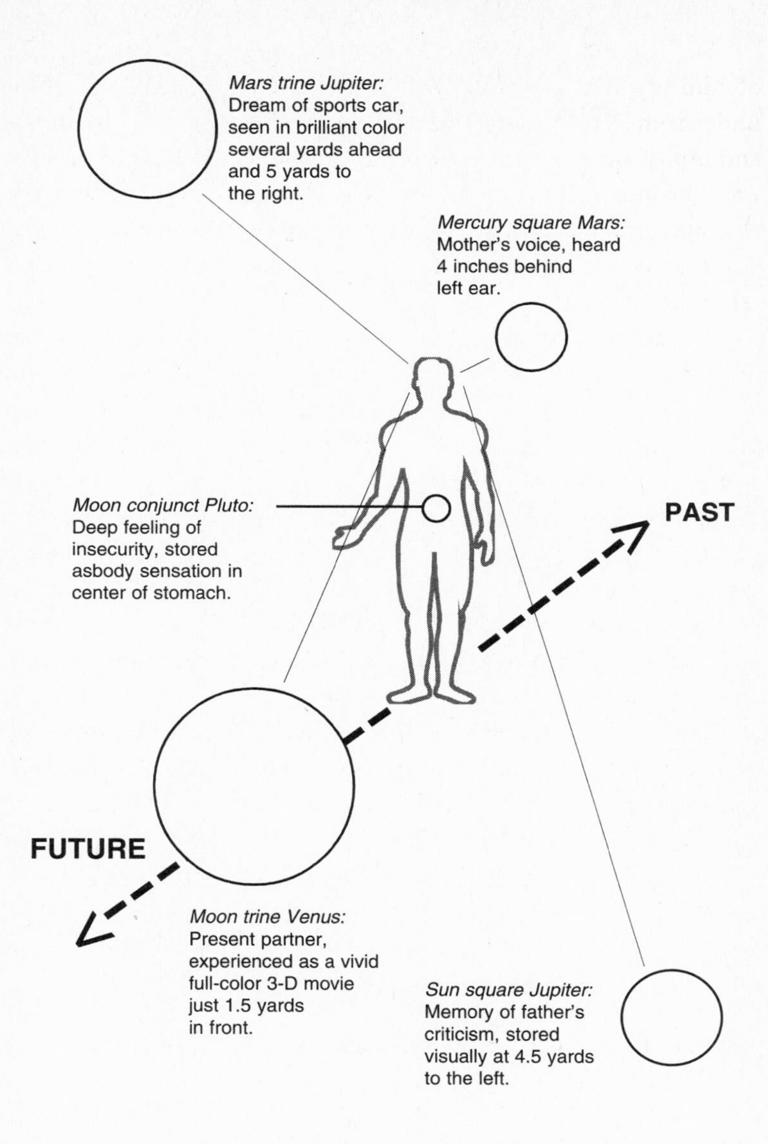

Fig. 39. Sensory storage coordinates. Experiences are stored as sensory impressions both within and outside of the body. As the client associates with these sensory impressions, the astrological energy associated with it is activated. When the coordinates of an impression are changed, the experience will change.

Astrologer: So, once a day, once a week, once a month . . . how often, precisely?

Client: Several times a week.

Astrologer: Was he a big man? [Working up to visual representation]

Client: No, well he seemed big, but he's quite small actually. [Client's eyes go up and to the left—he is now accessing father visually]

Astrologer: And, going back now to a violent episode, can you *see him striking you*?

Client: ... he used his belt.

Astrologer: So, describe the scene . . . he is standing there . . . you are . . . ?

[Client now describes in more detail]

Astrologer: So it really hurts, or what? [Working on physical representation]

Client: I simply turn off my body. I simply refuse to let it show.

Astrologer: So you're kind of deadened, physically, at this moment?

As he is reliving the sensory impressions in this interchange, the client is not going to be feeling at all good. For this kind of sensory reassociation exercise to be relevant in the consultation, the astrologer must be clear about the purpose. It is to reignite memories and decisions from the past that are having negative consequences at the present time. If, for example, the person cannot understand why his sex life is curiously flat or why he has cruel impulses, the accessing of memory will trigger a string of associations that can liberate him from the tyranny of hate, anger, or fear. Once the feelings have been accessed, it is important to move on to positive, transformative work.

If the square had been from Saturn in Aries to Mars in Cancer, then the interchange would take a different turn, because of the importance of the water element. Difficult memories stored in water signs are very dominating in regard to atmosphere, but far more challenging to access. Unless you have excellent rapport with the client, he will simply refuse to open up. When describing painful feelings to strangers (and, for water signs, people can remain strangers until they feel secure with them), the client will tend to look up and rationalize, rather than going down and inward to access body emotion. The exchange might go like this:

Astrologer: So, describe the scene . . . he is standing there . . . you are . . . ?

[Client now describes in more detail.]

Astrologer: So, what kind of *feelings* arise in you when this happens?

Client: Oh. I get on all right with my dad now. [Resolutely not looking down and accessing emotion]

Astrologer: No, wait . . . you're bending over, your dad's angry . . . you must be upset . . . or what? [Speaking slowly, gesturing down at client's body]

Client: I'm furious, actually.

Astrologer: So, is that a feeling you get in your legs, your arms, your head, your stomach . . . ?

Client: It's like my stomach is churning.

Astrologer: Like it's hot, or cold or acidic . . . or what? [Anchoring the sensory signals of the fury]

In this case, the questioning technique is geared toward liberating the power of the water/fire combination and its consequences in his everyday life. While there is every possibility that the astrologer already knows that Mars in Cancer square Saturn in Aries is going to show a lot of rage, it is better not to rationalize or interpret, but rather to allow the client to access the root energy himself in a process of self-discovery.

Staying with the Energy

Going back to the example of the Moon-Pluto client (figure 38), her anxiety could be alleviated by changing the way she represents the sound she hears. It is possible to guide clients into a more empowering sensory representation using whatever element you are working with. This can really help, but it is not a cure. To access transformative power, both client and astrologer need to stay with the energy. No astrological aspect is in itself bad. The core energy has no values attached to it; it is pure. It only seems traumatic or terrifying when there is the perceived threat of death to a part of the self. Yet, the part of the self that will die is actually the unhealthy manifestation of the original core energy. The energy itself cannot die, it can only transform. And it will transform if the client just lets go. Let's take the Moon-Pluto dialogue and guide it in another direction altogether:

Client: Yes, it's very loud, kind of staccato, the words are coming very fast. I'm frightened. She seems mad.

Astrologer: And this fear you feel . . . where in the body is it situated?

Client: My chest [disjointed breathing] . . . my solar plexus, really.

Astrologer: In your solar plexus like a pressure, heaviness, emptiness . . . what?

Client: It's just empty. It's a void.

Astrologer: How does that feel, that void?

Client: Empty. Endless.

Astrologer: And just staying with that empty, endless sensation . . . how is that?

[Trying to give the core energy more time to build up.]

Client: It's a black hole . . . it drains all my energy.

Astrologer: Like a vortex?

Client: Yeah. [Getting curious] It's immensely powerful.

You cannot know where this will go, but at this stage, it's just a question of not getting in the way with anything you say. No interpretation, no judgment, just suggestions to enhance the strength of the feeling. Generally, the client has to pass a painful or traumatic or even dangerous threshold to get into contact with the energy, which contains its own transformation. You would not want to do this with a client who was not mentally robust, and neither would you do it without some therapeutic training.

Staying with the energy demands both that the astrologer has confidence and experience with therapy and that clients are willing and able to work on a deep, inner energy that has been disrupting the quality of their life. Reaching the core energy is accomplished by evoking sensory experiences and stilling the rational mind. Every rational comment draws the client out of this trancelike state, while every evocative question induces a stronger state. When working with positive aspects, this is pleasurable; when working with negative aspects, it is painful. The client has to be willing to go over the threshold of psychological pain to induce transformation.

On this threshold, there will be considerable physical discomfort. Surprisingly often, this discomfort will be just where you would expect it, astrologically speaking. When the predominant energy is with the Sun or Leo, then the energy threshold is in the heart area; when with the Moon or Cancer, around the stomach; Venus or Taurus, in the throat area; Mars or Aries, in the head; Mercury or Virgo, in the solar plexus. These signs and planets are the most straightforward. Often, with Mercury-Gemini, the threshold is concerned with the intensity of inner voices; with Mars-Scorpio, the abdomen and groin. The Jupiter-Sagittarius threshold is often connected with existential issues and bodily sensations of expansion (with fear of a world with no limits), while Saturn-Capricorn can be connected to the physical sensation of oppressive weight. Uranus-Aquarius is often associated with the extremities of the body, especially lack of feeling in the fingers and toes; with this combination there is often an alienating disassociation from the body. The Neptune-Pisces threshold is related to loss of boundaries and the sensation of merging; subject and object become indistinguishable. Physically, this brings a sadness, which usually is experienced as a physical sensation, for example, in the throat area.

Working on the Threshold

You cannot count on the bodily sensation necessarily being associated with the astrological correspondence, but it happens often. It is important to let the client tell you where the sensations are, rather than you suggesting it to them. When you and the client have reached the threshold, several things can happen. Generally, the energy is so strong that it bears a transformative impulse of its own. It is as if the client has spent his or her whole life avoiding this energy, but when it is fully experienced it is no longer the terrifying gorgon it was once thought to be.

The following excerpt of a consultation shows the transformative power of Pluto (see figure 40, p. 278). The client in question—with a Sun-Pluto conjunction powerfully placed in the 8th house, emphasizing the potential for existential anxiety—suffered from a kind of agoraphobia, or irrational fears that prevented her from getting out and about. The astrologer guided her toward the energy threshold of Sun-Pluto in the following manner:

Astrologer: So when you're out in the street, you feel anxious?

Client: Yes, I just want to get inside as soon as possible.

Astrologer: What would happen if you just stayed outside?

Client: [Already getting nervous] I'd have a panic attack. I'd have to get home.

Astrologer: So this panic feeling . . . is that in the head, the stomach, or where? [Studiously avoiding a leading question, because with the Sun-Pluto in Leo, the feeling is in all probability . . .]

Client: Around here [pointing to her chest].

Astrologer: On the surface of the chest, or inside around the heart, or what?

Client: Around the heart.

Astrologer: Like a pressure, or pain, or vacuum? [Pluto often gives the sensation of vacuum.]

Client: [Feeling really uncomfortable] It's a dull pain. A deep emptiness . . . there's nothing there.

Astrologer: Like a black hole?

Client: [Interested] Yes! It sucks all my energy.

And that, of course, is exactly what a Sun-Pluto conjunction can appear to do. For the client, this is the basic energy on which identity is built. Her subsequent experiences spring from this energy and the ways she has learned to compensate for it. Yet, in the last analysis, her experience of the black hole is just an interpretation. Being sucked into it seems like dying. Perhaps she has residual memories of this happening, as the 8th house often brings a person into contact with death, rebirth, and karma. Dying always implies rebirth, so:

Astrologer: Even though it has felt unpleasant up to now, would you mind staying with that feeling of emptiness for a moment?

Client: [Reluctantly] OK.

Astrologer: Just allow it to get stronger. [Some time goes by.] This is the kind of feeling you get when you're out, right?

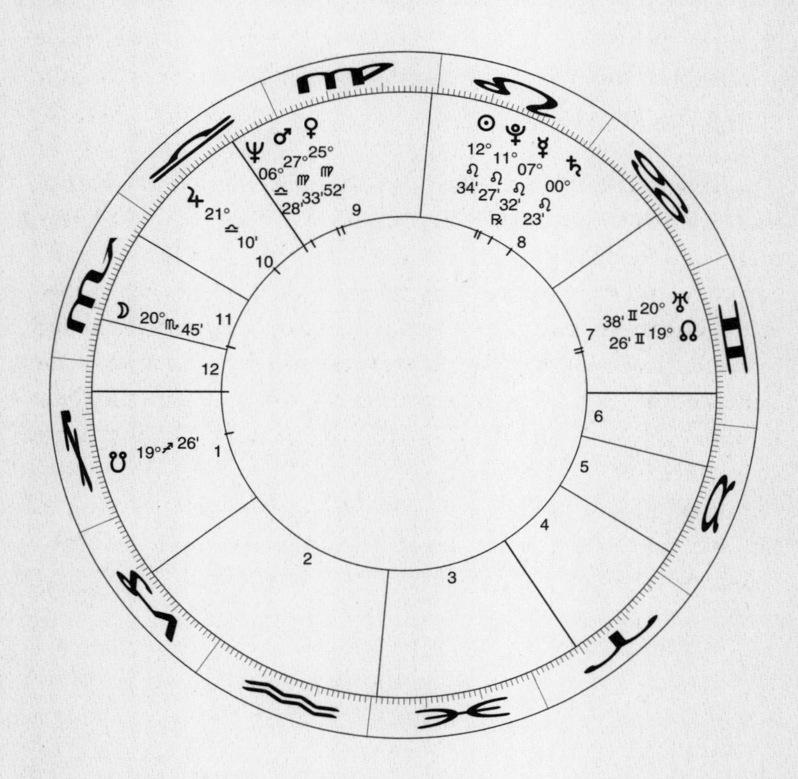

Fig. 40. Lene V. The Sun-Pluto conjunction in the 8th house evoked agoraphobic anxiety in this client, which she experienced as a "black hole" in her heart area.

Client: Yes.

Astrologer: [Using the metaphor] Did you know a black hole is seen to be black because it is so powerful that even light is sucked into it and cannot escape? A black hole is in fact full of intense light that cannot get out.

Client: Hmmm.

Astrologer: So all that energy that's been sucked in . . . it's very intense . . . feel the light in there.

Here the client was easily able to channel an intense energy that completely transformed the sensation of losing Self in emptiness to having an indestructible, luminous core of being. Pluto bears the energy of transformation within itself.

Each outer planet can be seen as providing unique transformative experiences. Jupiter, when allowed to expand without limits, merges consciousness with space, giving an experience of God. Saturn, with its immense concentrating energy, evokes an empowering consciousness of selfhood, destiny, responsibility, and strength. Uranus brings light energy, electrifying every cell of the body, and the light of awareness dawns. Neptune brings the client over the threshold of sadness, pain, and suffering to a timeless acceptance of the human condition. Pluto takes the client through the threshold of death, into rebirth; the old husk of self is burned up, and the Phoenix of newborn consciousness arises. All these transformations require courage.

16 Other Intervention Techniques—Secret Therapy, Metaphor, and Direct Feedback

A great deal of our life is private, or secret. Not many of us are inclined to reveal our innermost thoughts and feelings about sexuality to just anyone. A good therapist or astrologer may well be able to create an atmosphere secure enough to enable the client to open up about these matters, although this openness tends to vary culturally. Scandinavians, for example, are more comfortable than the English in the area of sexuality. Dysfunction in this area can be very damaging to a relationship. Considering that many problems are exacerbated by an inability to talk about them, it definitely pays to deal with sexual dysfunction in the consultation if it seems to be a key issue—which it would be with aspects from Mars to Neptune, Saturn, and Pluto in particular. It is surprising how many people actually do not mind going into detail when it comes to the crunch. However, it is also possible to work on private issues in a discrete way, and this can be very useful when working with sexuality. In a way, it is not necessary to know the facts concerning the manifestation of any core energy; it is only necessary to have the client access the sensory signals and

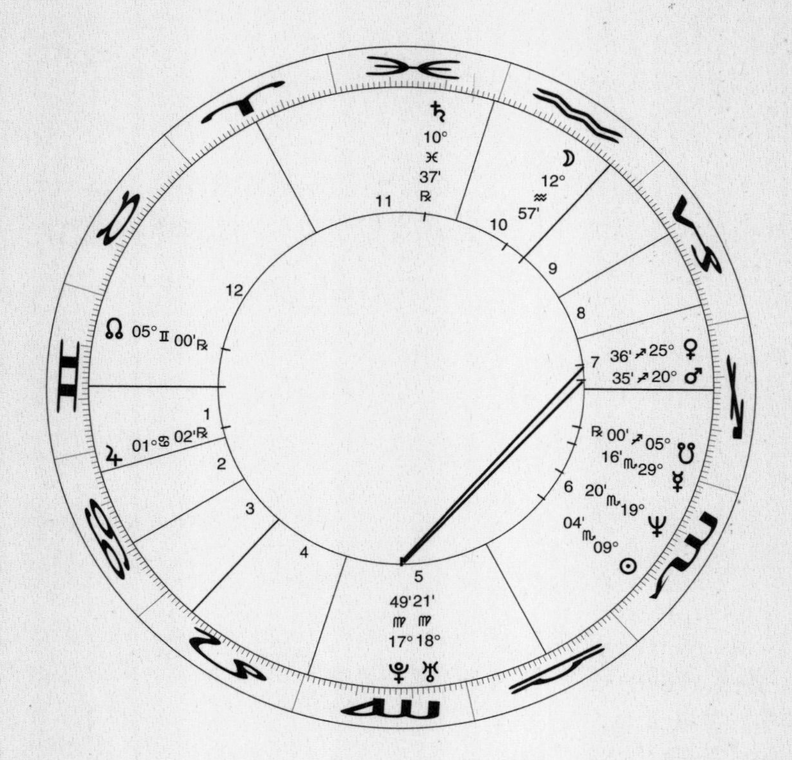

Fig. 41. Anette G. The powerful Mars-Venus square Uranus-Pluto reflects a sex drive that the client may prefer to keep secret.

work with them. In this way, you are just working with energy rather than content.

In the following example (figure 41)—which came up during an astrology course—it is clear that matters of relationships and sexuality are completely central to the chart, because of the involvement of Mars and Venus square Uranus-Pluto. The woman in question made a remark belittling her partner for never taking the initiative. Women with Mars-Pluto aspects will almost invariably complain that they themselves end up taking the initiative, especially sexually. So:

Astrologer: And when your partner takes the initiative, sexually, for example, how do you react?

Client: Well, he's so nervous and hesitant . . . I wish he'd just *do* it!

Astrologer: Isn't it true that when he "just does it" you start directing him as to what he should do?

Client: Yes, but it's because he's not doing it right!

Astrologer: Well that's the Mars-Pluto dilemma, isn't it? He's damned if he does, and damned if he doesn't.

[Mars-Pluto is a very "grateful" aspect; it's just dying to transform. In this public forum, it is still possible to work with intimacy while respecting the privacy of the student.]

Astrologer: I don't want you to say anything, but can you conjure up a picture of this scenario, just at the point where you take control?

Client: [Nods]

Astrologer: OK, now I know you would like him to take the initiative, so working on this inner film, run through a scenario in which you allow him to, paying particular attention to any anxiety you have in relinquishing control and the strategy you use for letting go. Just run the film through your head and let me know when you're finished. [With Mars-Venus in Sagittarius—a fire sign—it's a sure bet that her capacity to make visual images is unparalleled.]

Client: [After about 30 seconds] OK, I'm finished.

[She has now constructed a visual scenario, which means that she is still dissociated.]

Astrologer: Fine. Now run through the scene again, hearing what you might say, and whatever he may say, or whatever sounds there are. Let me know when you are finished.

Client: OK, I'm finished.

[Auditory scenario constructed]

Astrologer: Now this time run through the scenario and imagine any physical or emotional sensations that you may be having. Let me know when you're finished.

Client: OK. Good.

Astrologer: So when's the next time you are your partner are going to be together?

Client: [smiling] Tonight!

Astrologer: OK, just run through the whole scenario again imagining it is tonight, paying particular attention to how you relinquish control and to the whole range of sensations.

Generally it's very effective to "train" people by imagining a situation in the near future and going through it several times, with the new sensory strategies the client has privately worked out. It would take less than a minute to run through the process each time in the imagination. In the above exchange, the client has tacitly accepted her responsibility for her partner's supposed lack of initiative, and she has opened up and practiced new sensory routes, which will enable her to act differently when the situation arises. And this has basically been done without words. Rather than making the effect less powerful, this direct work with the senses makes the result even more effective, because so little is rationalized.

Metaphors and Stories

It is spellbinding to register clients' use of vocabulary. Far better than any astrological textbook, they supply verbal constructs that totally encapsulate astrological meaning. This choice of words will often be unique to them and their personal planetary combinations, and when they supply you with a living metaphor for their life, you're wise to pick up the terminology and feed it back. This instigates powerful mental rapport, which I discussed in the chapter on communication.

For example, a client with Moon in Gemini in the 9th, square Jupiter in Pisces, who was an only child (but she had, typically, "two" mothers—her mother had a twin sister), mentioned that on her mother's early death, she felt like "a helium balloon." A remarkable image for the 9th house airy Moon, "inflated" by Jupiter. When such an apt image is used, it is a sure route to a central part of the character, and it's important to hang on to it:

Astrologer: So you felt *cut adrift*, or what?

Client: Well, I was an only child, so it was as if I lost all connection. [Another Gemini word]

Astrologer: So you must have been *bursting* to talk to someone but felt quite alone?

Client: That's so true. I ended up reading so many books trying to discover some meaning in it all.

Astrologer: So then you must feel you've gained *a tremendous view* over these matters?

Using the balloon metaphor—"adrift," "bursting," "view"— makes the client feel understood on a profound level and facilitates in-depth work.

Metaphor is a huge subject, and there is a lot of literature about it.[11] One of the great advantages of using metaphors is that the conscious and rational mind of the client is flummoxed; the client does not necessarily know where you are going with a line of inquiry. As all deep change happens when the rational mind is quieted, this can be an advantage. There are many circumstances in which it is difficult to confront an issue head on and when the use of metaphor allows you to get around it. This can be when a person is oversensitive and defensive or too clever or too enlightened to really take seriously what you want to say about them.

In the astrological consultation this kind of circumspect approach can be rather simply done by telling stories. These stories can be true stories about yourself or another person, or stories you just make up.

One client was quite a well-known course leader, who, with a Sun-Uranus trine was very skilled at aiding personal development in groups. His partner had left him without warning, but he made it clear during the consultation that this was not his fault. However, aspects from Venus and Mars to Pluto in his chart told a different story, indicating that a less conscious part of his character was dominating, controlling, and jealous. One very disarming way of broaching this behavior is to claim it for yourself (if you have something similar):

> **Astrologer:** Oh! I've got something like this pattern, too! You know, I used to be so anxious that my partner would be unfaithful that I used to organize everything for her so that I'd know exactly where she was. At parties I'd get really upset if she talked too intimately with anyone. I know it was completely irrational, but I still could not prevent the anxiety. We used to have exhausting rows, and I ended up being hard and cold until she did what I wanted. It was terrible, really; I browbeat her. [This is a rough description of how Mars-Venus might operate in aspect to Pluto.]

It is just as effective to tell the story about yourself—if not more so—as to tell him directly that this might be the kind of behavior he

☉	Heroes Spotlights; being on stage. Kings, queens and royalty. Suns and stars; light givers. The beating heart, pumping life-giving blood. Leaders and their followers.	♃	Great thinkers and mentors who are intellectually respected. The expansion of the universe. The Horn of Plenty, seashells and objects that create expansion. Mountains with tremendous views.
☽	Belonging and attachment—to the womb, Mother Earth, home and hearth. Umbilical cords. Suckling, milk, everything that nourishes. Fertile valleys, rounded hills. Photos and family.	♄	Pressure that transforms coal into diamonds. Muscles that become strong through training. Walls that block, but protect. The trials of Job. Watches; grandfather clocks.
☿	Everywhere from whence messages arrive and depart. Letters and letterboxes. Mobility: bikes, buses, skates. Pen and pape News. Schools and learning.	♅	Explosions and their sudden transformative effect. Short-circuits and stress overloads. The exhilaration of jet aircraft, airports, and places with international atmosphere.
♀	The sweetness of romance and lovers. The pleasures of perfumes, sweets, a bouquet of flowers. An art gallery: the act of evaluating. Creating and enjoying music and harmony. "I'm worth it."	♆	The sea and its immensity. Not acting when enshrouded in fog. Swamps and morasses, and how to avoid them. Inner, spiritual light that even in death cannot be lost.
♂	Wars, battles, duels, and competitions. Iron and steel and the myriad of tools and weapons they compose. How strength is enhanced by victory and cutting through.	♇	Volcanoes, which bring destruction, yet create fertility. Black holes, which suck in light, but have the power to create universes. Abysses, which, when jumped into, are not so deep. Gorgons, which can be defeated when mirrored.

Fig. 42. Planetary metaphors. There are innumerable stories to construct around planetary effects so that awareness of their meaning and influence can be evoked.

has and that has made his partner leave without warning. A story like this goes straight into the unconscious, which sits up and takes note. Having set the scene, it is then possible to bring up solutions.

> **Astrologer:** Fortunately, I met someone who was able to give me some very good advice. He said to me [leaning forward], "Listen! Your anxiety about your partner stems from an inner fear from a much earlier period. Don't bother your partner with it. It's simply counterproductive. The more you allow anxiety to dominate, the more likely your partner is to leave. The more relaxed and democratic you are, the more she will want to stay with you." I can tell you he made a profound impact on me, and my relationship immediately improved.

The beauty of this method is that the client can take everything in without losing face. As a respected course leader in the New Age field, this man was not particularly inclined to reveal his weaknesses to me. It might seem unprofessional. The reported advice in the quote is actually said directly to the client and perceived by his unconscious as such. If you do not wish to use yourself as an example, you can also tell stories about other "clients": "You know I had a client who had a very similar pattern to this . . . "

The general idea is to describe a scenario as close as possible to what the client might be experiencing, describe the negative consequences, and then describe a brilliant solution. This is a particularly good method when talking to men about their sexuality, as male clients tend to be very uncomfortable in this area, especially with male astrologers.

Metaphors can be used proactively to help clients relate to change. The table in figure 42 can give you some idea of possibilities.

Direct Feedback

Some consultations seem to be crippled by a deadened atmosphere in which, no matter how completely you explain an influence, there seems to be no progress, no insights, no change. It is often the case that if you feel boredom, tiredness, anger, or frustration, the client probably evokes these feelings in others. By trusting your own feelings in this area, and commenting on them, there is often the possibility of a breakthrough. The assumption here is that because you feel something, the client may be the cause. It's a dangerous assumption, but acting on it can produce dramatic change in the progress of the consultation.

One client was describing all her troubles at work, and I found myself getting more and more irritated with her attitude, her voice, and general behavior. I noted her Moon-Mars conjunction in Aries in the 6th, and began to get a sense of what her coworkers were going through. I naturally spent some time trying to get over these feelings of irritation, but the consultation was going nowhere, so I thought it was worth risking something. I interrupted her, and explained how I had begun to feel more and more irritated by her. She was shocked and angered and, suddenly losing her temper, she got up and made as if to leave. I was surprised she reacted so violently, but let out an inspired comment: "I bet you've suddenly gotten up and left most of your jobs in this way, right?" It turned out she had left every job she had ever had by getting up and leaving without warning. She sat down; we subsequently had a great consultation, and she became a regular client.

It requires a certain amount of chutzpah to confront the client with your personal reactions to the client, but it brings results. You have to do it with sensitivity and, perhaps, as a last resort. It depends, too, on your character. It works when you function as an unbiased barometer, but not when your opinions and tastes are dominant.

Nelson Mandela used a similar intervention to turn around his talks with the right-wing white Nationalist leaders of South Africa in the early 1990s. Negotiations had been dragging on for months, and the major sticking point was the fear the Nationalists had that the black African leaders of the ANC were going to be unduly dominated by white Communists in the organization. Time and time again, Mandela assured them that this was not the case. Finally, in frustration he said, "You gentlemen consider yourselves intelligent, do you not? You consider yourselves forceful and persuasive, do you not? Well, there are four of you and only one of me, and you cannot control me or get me to change my mind. What makes you think the communists can succeed, where you have failed?"[12]

Here, Mandela used the evidence of his own body and mind, which those opposite him could see and hear, as a living testimony of truth. In the same way, you, as the astrologer, can present the client with your own reactions as evidence of the effect of the client. It's an invaluable opportunity for clients to receive unbiased testimony as to the impression they really give other people.

Synastry Consultations 17

In my first years of practice as an astrologer, I tended to dissuade couples from visiting me together. I felt it necessary to use the whole of the consultation period to go into depth with each individual in privacy in order to get to the bottom of behavioral patterns. Perhaps there would be a later consultation when both could come together. However, for a couple who wished to go in depth into their relationship, this could be an expensive option. I now take every opportunity to encourage couples to come together, simply because this can be the most effective form of consultation—also from an individual point of view. The only prerequisite is that they agree that *the relationship itself* will be the major subject of the consultation.

Corroboration of Behavior

There is one clear advantage of dealing with people as couples: the consequences of one person's behavior can be immediately confirmed or denied by the partner. When dealing with a single person in the consultation, you cannot really know how they are; all you have to go on are their claims and your subjective assessment of them. It's much easier when you have another person to back up your astrological hunches. Using the partner in this way is somewhat fraught with difficulty, as it is immediately confrontational, but if you have the nerve, the rewards are rich.

To give you a simple example, the astrologer may wish to describe the effect of the man's Mars-Jupiter square. It could go something like this:

Astrologer: [To man] This influence can lead to problems in relationships because of dominating intellectual behavior . . . you know, being a bit of a bully, or always insisting you're right.

Man: Well, I do have strong convictions, if that's what you mean.

Astrologer: Not exactly. [Turning to woman] Do you find he tends to pontificate and is not receptive to your own opinions?

Woman: Well yes, he *always* wants to have the last word.

The invaluable nature of this exchange is that the woman has probably repeatedly told the man how dominating he can be. With his Mars-Jupiter square, he has not been able to accept that he may be in the wrong. Now, out of the blue, the horoscope has independently confirmed this behavior. Once unwanted behavior has been identified in this way, it's important to follow through, to really anchor the new awareness:

Astrologer: [To woman] How does it feel when your husband insists on being right in this way?

Woman: It makes me feel a little stupid and intimidated.

Astrologer: [Turning to man] Do you want her to feel that?

Man: No, I certainly don't.

Astrologer: [Turning to woman] Can you tell him now what you would like him to do, and what not to do, when you discuss things?

During the first stage of the interchange, all communication goes via the astrologer, but it is effective at the end to make each person speak directly to the other. For them, it can feel that they are being listened to for the first time.

Reconciling Different Communication Systems

It is often the case that partners use quite different representation systems when they communicate. In other words, the horoscopes indicate that the man has a predominance in one set of elements and the woman in another, so that one prefers to send emotional and physical signals, for example, while the other likes to communicate and discuss. Love may bring a couple together, and children bind them, but in communication, they can be worlds apart.

The following couple participated as volunteers in a synastry course, and they were open and frank enough to admit to many differences and disagreements. They had one child and had openly discussed separation. Each of them had radically different ways of communicating, evidenced by very different Mercury-Gemini influences. The wife, Marie, had a specific complaint that when the husband, Johan, came home, he brusquely ignored their child who came to meet him at the door, hardly said a word to Marie, and hid himself away behind a newspaper on the sofa. Marie would then start arguing with him because he was ignoring everyone, and the whole atmosphere would be ruined. Johan, on the other hand, felt that Marie did not really listen to or understand him. Marie, sitting erect in her chair eloquently and convincingly claimed that she did listen to him, while Johan sat slouched in his chair, silent, arms protectively folded in front of him.

This was the meeting of two representation systems—the man with water influences, represented by the defensive Mercury-

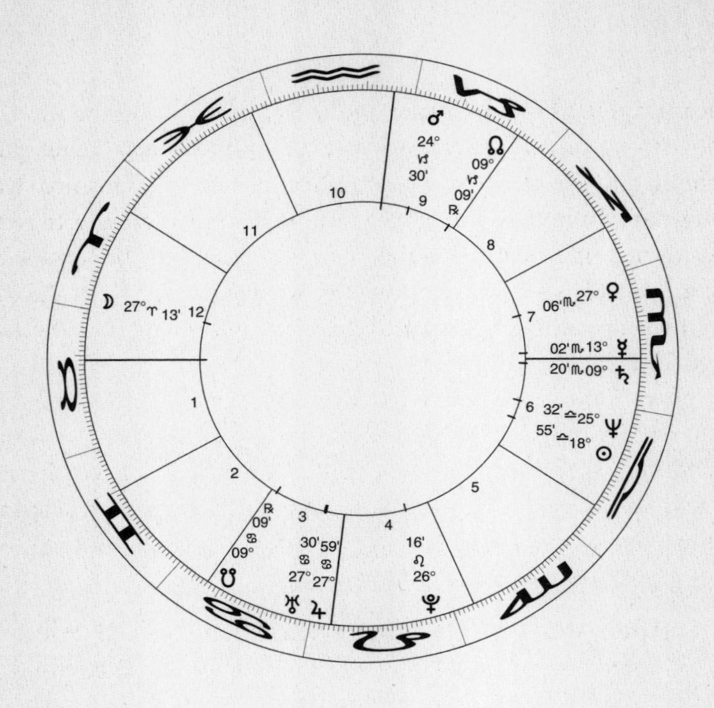

♂ 24° ♑ 30'
☊ 09° ♑ 09' ℞
☽ 27° ♈ 13'
06' ♏ 27° ♀
02' ♏ 13° ☿
20' ♏ 09° ♄
32' ♎ 25° ♆
55' ♎ 18° ☉
℞ 09' ♋ 09° ☋
30' ♋ 27° ♅
59' ♋ 27° ♃
16' ♌ 26° ♇
10
11
12
1
2
3
4
5
6
7
8
9

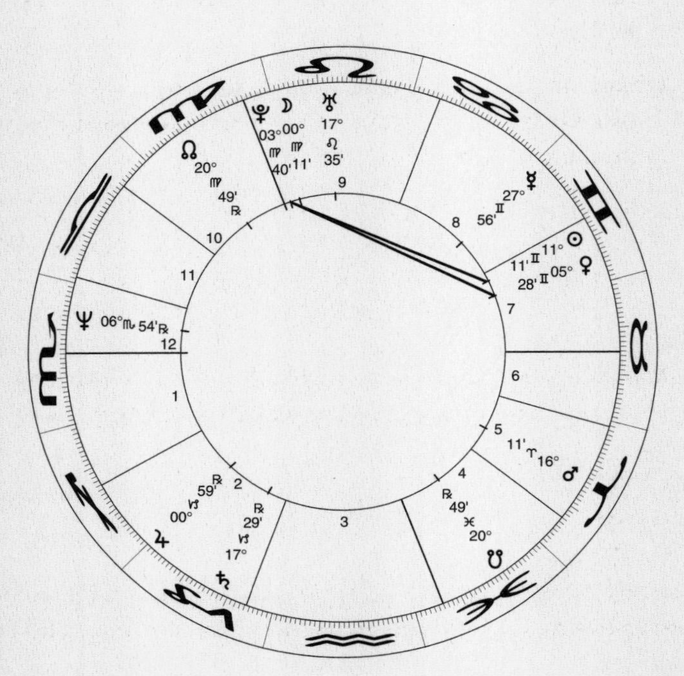

♇ 03° ♍ 40'
☽ 00° ♍ 11'
♅ 17° ♌ 35'
☊ 20° ♍ 49' ℞
27° ♊ 56' ☿
11' ♊ 11° ☉
28' ♊ 05° ♀
♆ 06° ♏ 54' ℞
11' ♈ 16° ♂
℞ 49' ♓ 20° ☋
℞ 59' ♑ 00° ♃
℞ 29' ♑ 17° ♄
9
8
7
6
5
4
3
2
1
12
11
10

Saturn conjunction in Scorpio on the Descendant, and the woman with strong air influences, represented by the chatty Venus-Sun conjunction in the 7th. Communication in relationships was important for both of them; they just did it very differently. In reality, Marie's lightweight, voluble nature bounced off the armor of Johan's deeply sensitive and receptive Mercury in Scorpio. She did not register the impression her words had on him, and he did not reveal how deep they went.

To facilitate communication in this instance, the technique is to adopt the water representation system for Johan, using emotive words and gestures, then re-present it in air, more intellectually, for Marie—translating, as it were, each person's language. This requires a proactive use of words, voice tone, and posture, so that both feel they are fully understood. This is how it was done:

Astrologer: [Turning to face Johan, matching with arms folded, posture slumped, speaking slowly and thoughtfully] How exactly do you feel, then . . . when you come in the door after work?

Johan: I'm just tired. [Pause, groping for words] I want to pick up my son and play, but . . . um . . . I just need a few minutes of quiet.

Astrologer: [Turning to Marie, now matching her by sitting erect, gesturing, and speaking brightly and more rapidly] So he's just come home, feels beat, needs a bit of a break, then he'll be ready to chat.

Marie: Yes, but it can't be too difficult for him to play and chat for a minute or two, *then* sit down and read.

Facing Page:

Top: Fig. 43. Johan.

Bottom: Fig. 44. Marie. This synastry consultation showed two people with very different element emphasis in terms of communication, one airy and the other watery.

Astrologer: [Turning to Johan, reverting to previous body language] So . . . um . . . Marie feels let down and . . . um . . . disappointed when you're silent.

Johan is reacting self-protectively in the only way he knows, and Marie has no tools for understanding this. The astrologer acts as an interpreter for them until each clearly understands the other. When Marie finally understood the grave communication difficulties that constantly challenged Johan in every relationship, she learned to respect them and adapt. When Johan realized that lack of communication triggered anxiety in Marie, indicated by the Moon-Pluto square to Sun-Venus, and that she interpreted his silence as a deeply critical comment on her self-worth, Johan discovered why Marie embarked on long nagging sessions. They made an agreement during this consultation that when Johan came home, she would give him a hug, pick up the child, and let him adjust for a short time before becoming talkative—instead of being offended by his noncommunicative manner. It worked well, and the last time I was in contact with them, many years later, a fourth child was on the way.

Of course, there is much more to synastry sessions, and the purpose of this section is just to point out the advantages of having the partner there to corroborate behavior. Synastry is a vast subject and merits a whole book, not least in connection with the interaspects between couples, shared aspects, and a myriad of other factors.

Points to Remember 18

Dialogues work better than monologues. While it is the job of the astrologer to unfold as much information as possible in a relatively short period of time, it is counterproductive to talk more than 70 percent of the time. Much talk leads to much exhaustion. If the client talks too much, the astrologer feels unproductive and frustrated, if the astrologer talks too much, the client loses concentration and gets overloaded. Dialogue insures maximum attention on both sides and makes the work of consultation a constant learning process for the astrologer.

The consultation is of mutual benefit. The most powerful consultations take place when client and astrologer interact and create a unified field. Energy for change is then present both for client and astrologer. Both are enriched by the interaction. It is a mutual growth and learning situation.

Establish your credentials at the beginning. First-time clients do not know what to expect and can be doubtful about astrology's effectiveness. Convince them right at the beginning by choosing one thing that you are sure of about their situation and telling them about it. Using the consultation chart makes this easy. Clients who are convinced the horoscope reveals their character, and that your are adept at reading it are much more open to psychological transformation processes. Clients who are skeptical about your skills will not trust you to help them change.

Getting it wrong is the first stage to getting it right. Clients are forgiving. They are surprised when you get something right, and not surprised when you get it wrong. It's all right to get something wrong, but it's not all right to ignore it and go on when told something is wrong. If you're wrong, stop and ask what's right—you are about to learn something new. Ask as many questions as necessary to fill out your knowledge so that you'll get it right next time.

Check it; ask questions. When you make a claim based on the chart, check it with the client to insure it's right. Clients are often too polite to correct you, but if you make an incorrect assumption early on, the whole consultation can be flawed. Rather than making statements, ask questions.

Ask intelligent questions. Questions should contain your proposal for the answer, rather than being open-ended. "Did you experience your mother as something of a martyr?" works, "How did you experience your mother?" doesn't. The client should sense that you know the answer and just need confirmation.

Describing the past has limited value. Bringing up the past alone muddies the well. Past behavior must be linked to present consequences, leading to new decisions about the future. Therefore, always relate past events to the present. The best consultations link past, present, and future when working on every significant pattern.

Blocked clients suggest astrologers who don't adapt. Communication and rapport can be established with anyone if the right matching takes place. Matching leads to understanding. When clients feel understood, they open. Powerful astrologers are able to put personal taste and preference aside to tune into, like, and respect even the most difficult clients they meet.

The client's experience of the world is not the world. In NLP there is a "presupposition" that states: "The map is not the territory." Every experience is transformed and mutated by our senses. This is normal. On top of this, we all warp the mutated sensory experience because of preconceptions, values, and personal problems. The client's reaction to the world is the exclusive result of his or her interpretation. Change the interpretation, and you change the world.

Use the client's map of reality. Whatever you think is right is only right for you. What's right for the client is what the client decides is right, not what you decide. When a client uses words, metaphors, and phrases to describe what they experience, there is a reason for each word, metaphor, and phrase. The world is—for the client—exactly as he or she thinks it is. Enter the client's world by using their map. The horoscope is the map.

The horoscope always works. Each word, phrase, sentence, conviction, and action is reflected by planetary dynamics in the horoscope. When listening to the client, locate the dynamic in the horoscope—it is always there. Finding this influence provides the key to change.

Know the 25 major aspects. Changes on a conscious level can be effected by accurate description of behavioral patterns and their consequences. Aspects between the inner, personal planets and the outer, collective planets are the key to different behavioral patterns. Make sure you know the unique manifestations of each of them.

Every astrological influence has a creative manifestation. Planets and their aspects can be empowering or disempowering. The toughest aspect has the greatest potential. The more difficult the configuration, the greater the potential for change, and the greater the respect due to the client.

Enroll the client. You can enroll clients into changing their map of reality by identifying behavior and the consequences of that behavior. When clients accept responsibility for consequences, they will want to change behavior. You can make suggestions of new behavior based on your experiences, but clients know best what changes are most practicable for them. New behavioral patterns should be arrived at in dialogue.

The horoscope shows how clients interpret sensory experience. Through the horoscope, you can identify sensory representations of problems and resources. Accessing these sensory representations empowers the client. Changing these representations changes the client's experience of the world.

When you change the client's world, others automatically change. No man is an island. Subtle energy links us to those close to us—especially family members. When the client successfully changes, then the family energy changes. A client helps family and loved ones best through inaugurating personal change and empowerment.

You must use energy to change energy. The deepest behavioral patterns are rooted in archetypal energy at the core of identity. Cosmetic changes and good intentions will have no affect on behavior. Permanent change can be wrought only by accessing the root energy and transforming it.

Clients have the resources they need to overcome problems. With very few exceptions, there is a balance between the strength and resources in the chart and the weaknesses and difficulties. When clients are in contact with their strengths, they have the power to transform their weaknesses. Therefore, work on evoking resources first and deal with problems afterward.

There is a positive intention behind every behavior. Whatever people do, they just want to be happy. Unconscious behavior has an unconscious positive intention. To change behavior, the original positive intention should be uncovered; if it has outlived its usefulness, it can be replaced.

The Universe has a positive intention. People who work with personal development believe in the idea of growth. Unless you think that you have achieved maximum personal growth, you will know that many people who have trodden the path of personal development before you have now reached higher spiritual levels. Love and kindness characterize spiritual development. Those further along the path are motivated to help us on the way. That's why the Universe has a positive intention.

Appendix

Meditation on the Four Elements

This appendix is aimed only at those interested in meditation, and perhaps only to a small proportion of them. Meditation practitioners have their own teachers and their own preferred meditation techniques, and they may not need or benefit from the following meditation. Those who do not practice meditation, but would like to, would be better served by first attending a class taught by a skilled teacher. But if you meditate and are interested in astrology, there is a chance that you would like this one. It is designed for those with astrological knowledge who might like to experiment with mental techniques to attain a more powerful appreciation of the four elements.

These elements are far more than matter locked in varying proportions into the manifestations of the material world. Rather than being represented by the liquids, solids, gases, and heat around us, they are nonmaterial energy states that inform matter, and around which all materialization take place. These energy states are inseparable from the consciousness that perceives them. The perception of each state induces a particular core psychological experience.

Astrological knowledge is in no way a prerequisite for these core experiences; it is just that a knowledge and experience of the four elements comes naturally with astrology, and this can be harnessed in the meditation. The whole point of this practice is to induce an experience of "body-mind" from which core experience can take place. This basically means feeling the body with such one-pointedness that body and mind fuse and are indistinguishable. This is a basic state from which true meditation can take place.

Although I refer during the meditation to the meaning and significance of the twelve signs, I do not intend to stimulate conceptual thought, ideas, theories, and speculation. Basically, while you are thinking, you are detached from your body and farthest from the body-mind experience, and when you are just experiencing or feeling sensations, then you are closest. It's difficult to both read and meditate at the same time, so you can download these instructions as a sound file from www.world-of-wisdom.com/meditation.htm—it's free of charge.

The meditation is divided into two parts: the first part is astrological and relates to the 12 signs; it is intended to completely relax the body and induce the body-mind state. The second part is simply concerned with the experience of the four elements and the fifth element of space. In fact, any preliminary breathing exercise or other body-relaxation meditation can effectively replace the astrological meditation. I developed the first meditation myself, based on some years of practice with body-relaxation meditation; the second is based on traditional Buddhist element meditation.

Inducing Body-Mind Consciousness: The 12 Signs

Begin by sitting comfortably. Hold your spine straight and erect, as if a force is gently lifting your head up from above. If you're sitting on the ground, it helps to have a firm and rather thick cushion supporting your buttocks and to have your legs crossed in front of you. Sitting on a chair is all right; the important thing is that your back is straight. Your head should be slightly inclined more forward than backward—this releases any tension in your neck. It's generally easier with your eyes closed but as if you are gazing at a spot on the floor one or two yards in front of you. Relax your shoulders, creating maximum distance between them and your ears. Imagine your shoulder blades flush with your back rather than protruding, and note how this creates more room in your chest and frees up your breathing. Breathe normally through your nose. Allow the tip of your tongue to rest on the roof of your mouth just behind your teeth. Take a few breaths now, with your attention solely focused on the air coming in and going out.

As you do this, you'll find thoughts appearing. When you notice this, let them go and bring your concentration back to your breathing. As you gain clarity during the meditation, you may be tempted to use this clarity to entertain a particular train of thought. This will not serve any useful function. While meditating, avoid following the well-trodden path of thoughts about past events and inviting thoughts about the future. All that matters is focusing on present sensory experience.

Earth: Capricorn, Taurus, and Virgo

Now, feel how your body is completely enclosed in its skin, thick in some parts, thin in others. Sense the boundary between inner and outer that your skin provides.

Now notice the rigidity and solidity of your body's skeletal structure.

Feel how skin and bone meet at the knee joint. It is here that there is balance between strength and humility. You need to be able to kneel in respect, but you cannot be weak in the knees. The joints are power nodes in the body. Feel the energy in your knee joints. Relax your knees, and just feel the sensation connected with the energy here for a couple of minutes.

Move your consciousness to your shoulders and your neck, checking that your spine is erect and your head light on top of your body and inclined slightly forward, so your neck is relaxed. See your shoulder blades flush with your back. Notice the strength of the top half of your back, and the vulnerability of your neck. This is where you put on the yoke and shoulder your burdens. Just allow yourself to feel the physical sensations in your neck and back while you relax in this area. You may notice quite strong emotions as you relax—especially at the front of your neck—but pass them by and simply concentrate on the feeling of relaxation for several minutes.

The next three signs are focused on the central area of the body, which is a key area for body-mind. As your consciousness rests in this area, your thoughts naturally become quieter and body awareness and power increase. Become aware of the sensations in the intestinal area inside the center of your lower body. This is where the most subtle nourishment is extracted and the dross expelled. You may experience rumblings here as you feel this area. Though sensations may be weak at first, allow your mind to rest here without distraction for a couple of minutes, and examine and relax into any sensory signals.

Water: Cancer, Scorpio, and Pisces

Become aware of the area around your breasts and stomach. This is where you give and receive nourishment. Notice how your sense of security and satisfaction is strongly linked to this area and how vulnerable you are through the need for food. Just experience any sensations here for a while—but without analyzing them—and completely relax your stomach area.

Move down to the lower abdomen and pelvic area, and relax. Many pleasurable and forbidden feelings are associated with this area. Allow any feelings and sensations to arise, without speculation, guilt, or judgment. Allow the feelings to center in your inner abdomen area about an inch below your navel. Just feel the energy at the center of the body here for a few minutes. Spend some time relaxing and focusing here—this is the powerhouse of your body.

Now concentrate on the physical sensations in your feet, examining your toes one by one and relaxing all the minute bones and muscles here. Farthest from the rational center of the brain, your feet are what connect you to the earth. You balance, lean forward, fall, and catch yourself with each foot in turn in the unconscious process of walking.

Just feel the energy in your feet, relax, and enjoy the sensation of the relaxation. The more you can feel your feet, the deeper your practice.

Fire: Aries, Leo, and Sagittarius

As you move your awareness from your feet to your head, there is a tendency for thoughts to gain ascendancy again. Note how strong the sense of self is as you focus on your head. Check that your spine is erect and your head slightly forward, as if pulled gently upward by an invisible force at the crown. Just concentrate on the physical sensations in your head. Feel your forehead and your jaw, and relax. Relax your eyeballs and feel the calmness of your gaze. Relax your mouth and the rest of your face. Without thought, just feel your head.

Moving your consciousness to your chest area and middle back, relax your rib cage and the muscles here. Notice how strong the sense of identity is at the center of your chest around your heart. There is a source of great strength here. Relax completely in this area and enjoy any well-being connected with this physical quietude for a couple of minutes.

Now become aware of the area between your hips and your knees. Feel the expanse of your thighs and their muscles, and the warmth of your blood suffusing them. Your thighs enable you to take great strides through life and give you mobility. Just concentrate on the warmth and energy.

Air: Libra, Aquarius, and Gemini

Focusing your attention on the small of your back, imagine your kidneys within your body maintaining a delicate internal chemical balance and purity. Become aware of the left and right side of your body: the two hemispheres of the brain; your two eyes, ears, and nostrils; your left and right arms and legs. Feel the poise and symmetry of your body, and be aware of balance and imbalance for several minutes.

Become aware of your blood circulation, bringing oxygen to the farthest extremities of your body through the smallest capillaries entering into your skin. Imagine, too, ethereal energies entering your skin from the outside, energizing you. Focusing now on your calves and particularly your ankles, feel warmth and energy in this region. Sense the energy flowing down your legs and through your ankles to your feet.

Now feel your hands, becoming aware of, and relaxing, each finger one by one. Relax your thumb, the palms of your hands, your wrists, and your forearms. Feel your breathing, becoming aware of your lungs filling with air and emptying again. Spend a few minutes concentrating on the tip of your nose, feeling the air coming in and going out. Whenever your thoughts start wandering, bring them back to the sensation of cool air coming in and warm air going out at the tip of your nose.

Inducing Body-Mind Consciousness: The Four Elements

Earth: Focus your consciousness deep inside your abdomen, at a point about two and a half inches below your navel, and from this point feel the energy of the earth element in your body. Become aware of everything that gives structure and solidity to the body: your bones, your skin, and inner cell structure. Imagine yellow light energy suffusing each cell of your body. There is a sense of heaviness, or of immobility, like a mountain. If distracted by thoughts, bring your consciousness back to your navel area. Your thoughts are stilled automatically when your consciousness is here. Allow awareness of your physicality to induce a sense of deep relaxation in your body. There is a sense of pleasure and satisfaction in the body energy. Allow yourself to take pleasure in the feeling of your body. If any area of your body feels particularly relaxed, explore and enjoy the feeling there.

Water: Become aware of the liquidity of your body: your tear ducts, saliva glands, bladder, blood, lymph system, and sweat glands. Feel how the liquid in your body gives a sense of cohesiveness and wholeness. Focus your attention inside your chest in your heart area, and imagine white light suffusing each cell of your body. Notice the great calm associated with the visualization of white light, and become aware of being centered in your self. The cohesiveness of water energy is associated with love. Allow yourself to become aware of this feeling as you keep your consciousness focused in the area of your heart.

Fire: Become aware of the warmth in your body. The surface of your skin is cool compared to the great heat generated within each cell of your body. Feel the relative warmth in different parts of your body. Focusing your consciousness in your throat area, imagine red light emanating from here and penetrating each body cell. The energy of fire in your body is connected to the sensation of wanting, or desire. At every moment, you are driven by wanting something, and this is the force driving you on in life. Feel the energy of desire that informs your sense of identity at any given moment.

Air: Within each cell of your body there is movement and space. This is the refined energy of air, which brings a sense of calm detachment. Focus your consciousness on the area between your eyes and imagine a blue light emanating from this point and suffusing each cell of your body. Associated with this energy is a sense of will. This will is the energy that drives us purposefully from one moment to the next, giving continuity in time. Feel your sense of will as you focus on the cellular energy inside your body.

Space: From the sense of movement and space connected with the air element, become aware of how matter and energy are interdependent and relative. What is within your body, and what is around it, is connected by the element of space, which unifies matter and energy in one continuum. Despite the relative density of your body, the space within it is vast, just as space outside it is vast. Become a part of that space.

As you end the meditation, concentrate on your breathing for a minute or so, and reconnect with your hands, your feet, and the rest of your body, before opening your eyes.

If you find that the meditation lasts too long for your busy schedule, it is more important to concentrate on the earth element than any of the other elements. We tend to live very much in our heads in the West, and we have lost the instinctive contact with the body that older, traditional societies had. The ability to rest in body feeling is linked to the ability to feel satisfaction and contentment in life, and the earth element enhances body feeling as well as a natural intuition that otherwise gets neutralized by the rational mind. The more your conscious energy is concentrated around the abdominal area, the more effective the meditation will be.

Notes

1. Nick Kollerstrom, *Astrochemistry: A Study of Metal-Planet Affinities* (London: Emergence Press, 1986).
2. See John G. Cramer, *Cold Fusion: Pro-fusion, and Con-fusion* (*www.npl.washington.edu/AV/altvw36.html).*
3. For a very succinct analysis of the philosophical basis of astrology, read Tai Situpa XII, *Relative World, Ultimate Mind* (Boston: Shambhala, 1992).
4. The clearest exposition of the concept of unity in duality is by Tarab Tulku, who holds courses on Buddhist philosophy and psychology throughout Europe. More information can be found on *www.suxess.net/tarab-institute/int/eng/tarabtulku.htm.*
5. See my book, *Doing Time on Planet Earth*, (Rockport, MA: Element, 1990) for a detailed examination of the consultation horoscope.
6. Most notably by Richard Bandler and John Grinder in *The Structure of Magic*. (Palo Alto, CA: Science and Behavior Books, 1975). Chapter 4, "Incantations for Growth and Potential" is dry but essential reading.
7. See C. G. Jung, *Collected Works,* vol. 16, § 543–548 for information on the medieval humors related to the senses.
8. My personal favorite is Robert Pelletier, *Planets in Aspect* (Atglen, PA: Schiffer Publishing, 1980). His ability to describe the subtle difference between sextiles, trines, quincunxes, and so forth is amazing.
9. Read, for example, Connirae and Steve Andreas, *Heart of the Mind* (Moab, Ohio: Real People Press, 1989). Chapter 7, "Phobias, Traumas and Abuse," gives a vivid demonstration of a safe way to deal with phobias.
10. See, for example, Tad James and Wyatt Woodsmall, *Time Line Therapy and the Basis of Personality* (Cupertino, CA: Meta Publications, 1988).
11. For more on metaphor, read David Gordon, *Therapeutic Metaphors* (Cupertino, CA: Meta Publications, 1978).
12. Nelson Mandela, *Long Walk to Freedom* (New York: Little, Brown & Co, 1994).

Bibliography

Astrology

Baigent, Michael, Nicholas Campion, and Charles Harvey. *Mundane Astrology.* Wellingborough, UK: Aquarian Press, 1984.

Costello, Darby. *Earth and Air*. London: The Centre for Psychological Astrology Press, 1999.

Costello, Darby. *Water and Fire*. London: The Centre for Psychological Astrology Press, 1998.

Duncan, Adrian. *Doing Time on Planet Earth*. Rockport, MA: Element, 1990.

Greene, Liz. *Saturn – A New Look at an Old Devil.* York Beach, Maine: Samuel Weiser, 1976.

Kollerstrom, Nick. *Astrochemistry: A Study of Metal-Planet Affinities*. London: Emergence Press, 1986.

Pelletier, Robert. *Planets in Aspect.* Atglen, PA: Schiffer Publishing, 1980.

Schulman, Martin. *Karmic Astrology – The Moon's Nodes and Reincarnation.* New York: Samuel Weiser, 1975.

Sullivan, Erin. *Where in the World.* London: The Centre for Psychological Astrology Press, 1999.

Buddhism

Chödrön, Pema. *The Wisdom of No Escape*. Boston: Shambhala, 1991.

Dhargyey, Geshe Ngawang. *Kalacakra Tantra*. Dharamsala, India: Library of Tibetan Works & Archives, 1985.

Guenther, H. V. *Tibetan Buddhism in Western Perspective.* Emeryville, CA: Dharma Publishing, 1977.

Guenther, H. V. and Leslie S. Kawamura. *Mind in Buddhist Psychology*. Berkeley: Dharma Publishing, 1975.

Tai Situpa XII. *Relative World, Ultimate Mind*. Boston: Shambhala, 1992.

NLP

Andreas, Connirae and Steve. *Change Your Mind – and Keep the Change.* Moab, UT: Real People Press, 1987.

———. *Heart of the Mind.* Moab, UT: Real People Press, 1989.

Bandler, Leslie Cameron. *Solutions.* San Rafael, CA: Future Pace Inc., 1985.

Bandler, Richard and John Grinder. *The Structure of Magic.* Palo Alto, CA: Science and Behavior Books, 1975.

Gordon, David. *Therapeutic Metaphors.* Cupertino, CA: Meta Publications, 1978.

James, Tad and Wyatt Woodsmall. *Time Line Therapy and the Basis of Personality.* Cupertino, CA: Meta Publications, 1988.

Other

Capra, Fritjof. *The Tao of Physics.* Boulder: Shambhala, 1975.

Mandela, Nelson. *Long Walk to Freedom.* New York: Little, Brown & Co., 1994.

Index